T0320074

Redesigning Environmental Valuation

NEW HORIZONS IN ENVIRONMENTAL ECONOMICS

Series Editors: Wallace E. Oates, *Professor of Economics, University of Maryland, College Park and University Fellow, Resources for the Future, USA and* Henk Folmer, *Professor of Research Methodology, Groningen University and Professor of General Economics, Wageningen University, The Netherlands*

This important series is designed to make a significant contribution to the development of the principles and practices of environmental economics. It includes both theoretical and empirical work. International in scope, it addresses issues of current and future concern in both East and West and in developed and developing countries.

The main purpose of the series is to create a forum for the publication of high quality work and to show how economic analysis can make a contribution to understanding and resolving the environmental problems confronting the world in the twenty-first century.

Recent titles in the series include:

Integrated Assessment and Management of Public Resources
Edited by Joseph C. Cooper, Federico Perali and Marcella Veronesi

Climate Change and the Economics of the World's Fisheries
Examples of Small Pelagic Stocks
Edited by Rögnvaldur Hannesson, Manuel Barange and Samuel F. Herrick Jr

The Theory and Practice of Environmental and Resource Economics
Essays in Honour of Karl-Gustaf Löfgren
Edited by Thomas Aronsson, Roger Axelsson and Runar Brännlund

The International Yearbook of Environmental and Resource Economics 2006/2007
A Survey of Current Issues
Edited by Tom Tietenberg and Henk Folmer

Choice Modelling and the Transfer of Environmental Values
Edited by John Rolfe and Jeff Bennett

The Impact of Climate Change on Regional Systems
A Comprehensive Analysis of California
Edited by Joel Smith and Robert Mendelsohn

Explorations in Environmental and Natural Resource Economics
Essays in Honor of Gardner M. Brown, Jr.
Edited by Robert Halvorsen and David Layton

Using Experimental Methods in Environmental and Resource Economics
Edited by John A. List

Economic Modelling of Climate Change and Energy Policies
Carlos de Miguel, Xavier Labandeira and Baltasar Manzano

The Economics of Global Environmental Change
International Cooperation for Sustainability
Edited by Mario Cogoy and Karl W. Steininger

Redesigning Environmental Valuation
Mixing Methods within Stated Preference Techniques
Neil A. Powe

Redesigning Environmental Valuation

Mixing Methods within Stated Preference Techniques

Neil A. Powe

Lecturer
School of Architecture, Planning and Landscape,
Newcastle University, UK

NEW HORIZONS IN ENVIRONMENTAL ECONOMICS

Edward Elgar
Cheltenham, UK • Northampton, MA, USA

Published by
Edward Elgar Publishing Limited
Glensanda House
Montpellier Parade
Cheltenham
Glos GL50 1UA
UK

Edward Elgar Publishing, Inc.
William Pratt House
9 Dewey Court
Northampton
Massachusetts 01060
USA

A catalogue record for this book
is available from the British Library

Library of Congress Cataloging in Publication Data
Powe, Neil A., 1969-
Redesigning environmental valuation : mixing methods within stated preference techniques / by Neil A. Powe.
p. cm. – (New horizons in environmental economics series)
Includes bibliographical references and index.
1. Environmental economics. 2. Valuation–Environmental aspects. 3. Environmental quality–Valuation. I. Title.
HC79.E5P675 2007
333–dc22

2006102433

ISBN 978 1 84542 279 0

Typeset by Manton Typesetters, Louth, Lincolnshire, UK
Printed and bound in Great Britain by MPG Books Ltd, Bodmin, Cornwall

Contents

Tables

Acknowledgments

A number of people have been very helpful in the process of producing this book and I wish to thank them for their efforts. These have included Douglas Macmillan for his help with the 'Market Stall' technique; Ian Bateman, Roy Brouwer, Guy Garrod, Nick Hanley, Paul McMahon, William Wadsworth and Ken Willis from whom I have learnt a great deal whilst co-working on mixed method valuation studies; Ricardo Scarpa for his encouragement; and Jan Wright for her invaluable help reading through the text and making a number of useful suggestions. Special thanks also go to Lena Sünnenberg and Gilla Sünnenberg.

1. Introduction

Interest in environmental conservation, biodiversity preservation and outdoor recreation has increased over the last few decades, with pressures on such natural resources for income generation following a similar upward trend. Consequently, conflicts have occurred between their destructive direct and associated uses[1] and an increasing desire for their preservation. Traditionally such difficulties have not been seen to occur, as, on the basis of financial returns, environmental amenities have little, if any, value. This absence of market prices has led to the stock of wetlands, forests and other natural assets being substantially diminished. Economists generally refer to this situation as market failure[2] where, through either the absence of a market or its inefficient operation, the total economic value to society is not considered within market operations.

Economists have devised a range of methods in order to include non-market goods within the analysis of costs and benefits. These techniques take advantage of the assumption that although no direct market for the goods considered exists, individuals are still willing to pay something for the benefits received. For example, surrogate market techniques examine the relationship between non-market goods and actual markets. Alternatively, stated preference approaches use questionnaire surveys to create market/referendum like situations within which good definition, means of payment and rules are outlined.

The contingent valuation (CV) method has provided the main focus of stated preference research and more recently choice experiments (CE) have been gaining in popularity. Stated preference methods provide the greatest potential of all non-market valuation approaches as their versatility enables them to be used to value many different types of goods and services. Indeed, stated preference methods have been widely applied for benefit estimation including: air visibility; clean water; woodland; future landscapes; transport safety; street lighting and heritage sites. Furthermore, unlike surrogate market approaches, stated preference methods can be used to estimate the full range of economic benefits received. However, some commentators have questioned the effectiveness of stated preference methods (Kahneman and Knetsch, 1992; Hausman, 1993; Diamond and Hausman, 1994; Boyle *et al.*, 1994). Despite these criticisms, cautious but positive assessments of the usefulness of stated preference methods have also come from many sources (Mitchell and Carson, 1989; Arrow *et al.*, 1993; Carson and Mitchell, 1995; Cummings and Taylor, 1999; Bennett and

Blamey, 2001). Given this controversy, there are many challenges faced when conducting stated preference surveys and the interpretation of the responses requires an understanding beyond neo-classical economics.

CHALLENGES FACED

Given the extent of criticism, no exhaustive categorization of challenges facing stated preference practitioners can be made. However, there are three recurring themes within the literature which are particularly pertinent to the issues considered when using qualitative methods, namely:

1. *Cognitive task faced by respondents*. Here it is important to understand the implications of respondents struggling to understand the scenarios presented and how they feel about the scenarios. It is also important to remember the novelty of the elicitation mechanism within which the respondents have to state their preference. The cognitive task is particularly great for the consideration of environmental scenarios, which are inherently complex and can be characterized in terms of uncertainty of outcomes, lack of clarity as to the best environmental outcome and difficulties of demarcation. The task of the researcher is to reduce respondent cognitive load whilst maintaining sufficient understanding so that meaningful responses can be elicited.
2. *Hypothetical nature of the transaction*. To economists the hypothetical nature of the choices made within stated preference methods is a key concern. As a general principle, it is crucial that the perceived linkages between response and policy-formation are sufficient for respondents to take the price seriously and put in the required effort to give meaningful answers. The hypothetical nature of the stated preference transactions, where the link between response within the questionnaire and payment may not feel as binding as other forms of transaction, is seen to be a key reason for respondents putting in insufficient effort. Indeed, study design needs to focus on providing a scenario that is incentive compatible providing thus truthful preference revelation.
3. *Communal nature of the scenarios considered*. The communal nature of environmental decisions means that the implications are shared by others and there may be moral and ethical issues involved. This communal nature usually requires the use of a collective payment vehicle which implies something about property rights and who else pays. These issues complicate the meaning of stated preference responses and respondent disagreement with the implicit principles of the scenario considered may lead, for example, to a high refusal rate or responses whose meaning is difficult to

interpret. The communal nature of the scenarios may mean that respondents are torn between responding as citizens or as consumers, where there may be incommensurability between the two. The challenge for practitioners is at least to gain an understanding of the extent to which these issues cause difficulties for the respondents and whether responses are consistent with economic theory.

USE OF QUALITATIVE METHODS

Given the challenges outlined above, it has become perhaps inevitable that there is a need to mix methods within stated preference surveys. Indeed, it has become conventional to use qualitative methods (in-depth individual interviews or focus groups) within questionnaire design. More recently, qualitative methods have also provided a wealth of information that has greatly enhanced our understanding of the meaning and acceptability of stated preferences, aiding analysis and helping to find a more appropriate role for valuations within policy decision making. More specifically, qualitative methods have been used to:

- gain an understanding of perceptions, categories and language used when considering stated preference questions;
- pilot a questionnaire or part of a questionnaire identifying any misperceptions in the scenario definition and related concerns;
- determine additional explanatory variables to be elicited within the main survey to aid interpretation of valuation responses;
- help explain post-survey any empirical regularities or anomalies encountered within the quantitative results; and
- consider the public acceptability of the approach as an aid to environmental decision-making.

These objectives can all be meaningfully achieved without any fundamental changes to the conventional stated preference approach.

Despite the improvements to questionnaire design achieved through the use of qualitative methods, it has been demonstrated that conventional stated preference research may still:

- make insufficient allowance for the cognitive limitations of the respondents;
- provide insufficient incentives and;
- make insufficient allowance for the communal nature of the scenarios and payment vehicles.

Although the use of group methods can do little to correct for the hypothetical nature of stated preference, it has been suggested that cognitive and communal issues can be better dealt with by further extending the role of the group-based approach. For example, group methods provide a permissive and non-threatening environment for value construction and follow-up questionnaire/meetings can also provide the opportunity for post-meeting discussion with friends/family and further research of the issues. Group-methods also enable the communal nature of environmental issues to be considered within a social environment and consensus to be reached on environmental values. Although alleviating the problems of preference construction and communal issues, such departures from the conventional approach may also have implications in terms of the form of the value estimates calculated and their use within decision making.

AIMS AND CONTENT

This book explores the extent to which the challenges of stated preference methods can be overcome through the use of mixed methodologies, particularly qualitative methods. More specifically, the book:

- introduces stated preference and qualitative methods;
- provides a step-by-step guide to the use of qualitative methods within environmental valuation;
- considers insights from qualitative methods into the meaning of stated preference responses and the applicability of the methods used; and
- explores how the role of group-based approaches can be extended beyond that of a complementary role to further improve environment valuation.

Part I provides an introduction to environmental issues and public consultation, as well as stated preference (contingent valuation and choice experiments) and qualitative methods. The first chapter in this section (Chapter 2) provides an overview of the complexity of the environmental scenarios decision makers are faced with and considers the various approaches available in terms of public consultation. Chapter 3 then introduces stated preference methods and the challenges faced in designing questionnaires that will provide meaningful valuations. The practicalities and pitfalls of using qualitative methods to improve the conventional valuation process and design are then considered (Chapter 4) in the form of a guide to their implementation.

Part II considers the contribution to stated preference from mixing methods. A detailed first chapter (Chapter 5) provides a framework for interpreting stated preference responses in term of economics, behavioural psychology and more social approaches. This chapter also provides an outline of the many insights gained from qualitative methods into the meaning of the valuation responses given and the applicability of stated preference approaches. Chapter 6 then reviews these issues in the context of scope sensitivity which has been the subject of a heated empirical debate within environmental economics literature and is crucial to the question of applicability of stated preference research.

The final Part looks at further extending the use of qualitative methods within environmental valuation research. Chapter 7 considers the use of group-based approaches to actively deal with the problem of preference construction. Chapter 8 explores the extent to which group-based methods can also aid difficulties caused by the communal nature of scenarios and payment vehicles used. These two chapters illustrate the many recent efforts that have been attempted in order to develop 'better' environmental valuation estimates. The results are promising but also illustrate the challenges of trying to mix methods within environment valuation.

The book concludes by evaluating stated preferences in the context of qualitative findings and provides a judgement as to the extent to which they can be improved through the use of mixing methods.

NOTES

1. For example, agriculture or housing development would be a direct use and sewage disposal an associated use.
2. See any basic environmental economics text such as Hanley *et al.* (2001).

PART I

Methods

Introduction

Part I of this book considers the methods which can be used within public consultation on environmental issues. Within this part, Chapter 2 provides an outline of the fundamental issues, including a discussion of the difficulties in gaining public preferences on environmental issues, the alternative methods that can be used and, within this context, considers the role of stated preference research. Chapter 3 provides a more detailed introduction to stated preference. The chapter begins with the necessary explanation of public goods, cost-benefit analysis and total economic value. The stated preference methods of contingent valuation and choice experiments are then introduced, before considering the essential elements of the hypothetical transactions that are developed (presentation, payment vehicle and elicitation method). Chapter 4 provides a detailed explanation of qualitative methods and begins by comparing qualitative to quantitative methods as well as group and individual qualitative methods. Having carefully outlined these principles, a step-by-step guide is given on how to conduct focus groups to improve the conventional valuation process and design. This detailed presentation is also very useful to illustrate key issues related to the more general use of qualitative methods within stated preference surveys.

2. Need for public consultation and stated preference

INTRODUCTION

A recurring theme throughout this book is the complexity and the problematic nature of consultation on environmental issues. For example, challenges are often encountered within consultation in terms of presentation to the time limited general public as well as the contrasting/pluralist nature of the values people hold. This complexity is further increased by the uncertainties and irreversibilities associated with environmental issues. This chapter provides an overview to this complexity and considers the various approaches available in terms of public consultation. Stated preference, the focus of this book, will then be introduced within this context.

CHARACTERIZING ENVIRONMENTAL GOODS AND SERVICES

Environmental issues are inherently complex and can be characterized in terms of uncertainty of outcomes, lack of clarity as to the best environmental outcome, difficulties of demarcation and pluralism of values.

There are many limitations and uncertainties within environmental science where dealing with imperfect information is perhaps the norm. Vatn and Bromley (1994) referred to this problem as functional transparency, where transparency occurs as the 'precise contribution of a functional element in the ecosystem is not known – indeed is probably unknowable – until it ceases to function' (p. 133). Given this uncertainty of outcomes and potential irreversibility of actions, translating scientific knowledge into public policy is difficult, with the precautionary principle providing an illustration of one way to deal with these inadequacies. Indeed, even with improved information on the environmental effects of actions, decisions within the scientific community are often unlikely to be clear cut. Consequently, where agreement is not reached between the many branches of environmental science there will be a need for priorities to be set. For example, wind turbines reduce carbon emissions however some of the most suitable locations may be near coastal bird habitats and decisions as to which environmental aspects to prioritize needs to be made. O'Neil (1997) further illustrates this decision dilemma with a wetland example; ornithologists tend to favour greater reed bed and open water but botanists would be concerned that this would destroy some of the most interesting plant communities. To the general public such uncertainty of outcomes and lack of scientific consensus as to the 'best' environmental outcome will be difficult to comprehend and, as such, requires there to be some sort of education process within public consultation and there needs to be an awareness of the cognitive ability and requirements on those consulted.

A further difficulty facing those wishing to increase public participation relates to the challenges of characterizing environmental goods and services in a way similar to that of private goods. There are two aspects to this: the public good nature; and the interrelated character of environmental issues. From an economic perspective, environmental issues to some extent have the public good characteristics of non-rivalry and non-excludability. A good is non-rival or indivisible when one individual can consume a unit of the good without detracting from the consumption of others. Non-excludable goods cannot be withheld from the enjoyment of others, where carbon fixing provides a good possessing both non-excludable and indivisible characteristics. As such consumption will involve others and public goods will be communal in character. This is particularly the case in terms of externalities, where harming behaviour, for example pollution, will affect the enjoyment of others. This is not usually the case for private goods. A further difference relates to the free rider problem. In order to alleviate this problem a communal solution is often required to both their provision and payment, which may be made compulsory following a consensus decision achieved through some form of public consultation.

The second issue relates to what Vatn and Bromley (1994) refer to as the 'composition problem', where they suggest a 'precise valuation requires a precisely demarcated object' (p. 137). As such, providing boundaries for environmental goods and services is difficult, as is the attaching of property rights. Understanding of ecology is based on a holistic view of nature where there is interdependence within ecosystems based on a long association between species. As such each part of the system may be equally as important to its whole. Sub-dividing eco-systems, or areas for the purpose of policy such as road construction is inherently difficult, with uncertainties as to the outcomes.

Finally, given the communal/social nature of environmental issues, they tend to represent a wider range of issues than consumer goods. Further to personal enjoyment for recreation and/or the environmental functions that ecosystems provide there may also be moral obligations/ethical principles associated with their provision. These may include perceived 'rights' of wildlife and/or feelings of 'doing our bit' for the environment. Given the communal nature of environmental issues and the difficulties of assigning property rights there may be differences of opinion between government/private interests and citizens regarding who holds the appropriate property rights. There may also be difficulties comparing the needs of current to future generations further complicating the decision making process. It has been argued that these pluralistic values cause difficulties in choosing an appropriate method of consultation (Vatn and Bromley, 1994; O'Neil, 1997). Of particular note has been the difference between the private interests (paying for environmental schemes and privately received benefits/externalities) and the more citizen type (moral concerns, or right and wrong), where the choice of consultation method will affect the particular emphasis on private and citizen type concerns (Sagoff, 1988).

EXPERTS AND REPRESENTATIVE DEMOCRACY

Given the complexities of environmental decisions, one approach could be to 'leave it to the experts'. It has been argued above that, in the case of scientists, it is often unlikely that a clear cut objective solution will exist in terms of what is best for the environment. Where there are difficulties general public input would be needed into determining the priorities, with scientists asked to consider how such priorities can be reached. However, often decisions are wider than that of a scientific debate regarding the best environmental solution. For instance, there are various ways that water can be supplied for household and industrial usage where there is a need to trade-off the quality of supply with environmental impact and price (Powe *et al.*, 2004b). The consideration of such alternatives has led to debates between scientists regarding the different environmental implications as well as a range of other interrelated non-environmental relationships. Through regulation or state ownership there is a role for elected representatives to make judgements based on the competing claims of self-proclaimed interest groups on either side of the argument (Vatn and Bromley, 1994). Indeed, research into most environmental decisions will reveal this process in action.

Continuing with the water supply example, it could be also argued that water consumers are very relevant to the decisions of water companies as the choices made will affect their supply, how much they pay, and any benefits they receive from their enjoyment of the environment. Furthermore, there may be imperfections in leaving decisions to a representative democracy where governments may pursue the interests of some particularly strong pressure group or the self-interest of the politicians and bureaucrats involved. Further to findings that expertise does not necessarily improve judgement (Payne *et al.*, 1992), the lack of a systematic and open evaluation procedure may mean environmental decisions do not reflect the preferences of society as a whole. Consequently, this has led to what Turner and Jones (1991) refer to as 'interrelated market and intervention failures'. For example, in terms of the loss of wetland habitat, Turner *et al.* (2003) suggest that although some decisions may have been made in the public interest (where the returns of alternative uses have been high), wetlands have frequently been lost to activities resulting in only limited benefits or, on occasion, even costs to society. Such failures, fuelled by a lack of public trust in both scientists and decision makers, have led to a need for greater public involvement within environmental decision making.

ALTERNATIVE FORMS OF CONSULTATION

There are various forms of consultation regularly used for considering projects affecting the environment, each of which raises important issues that will be returned to later in this book. Chess and Purcell (1999) suggest the quality of participation can be assessed in terms of both the process involved (fairness, representativeness, breadth and depth of discussion, degree of independence), and the outcomes of this process (inform policy, achieve consensus, make better accepted decisions). This will clearly depend on the perspective of those evaluating the process. In their literature review of public participation on the environment, Chess and Purcell (1999) found the success of public participation effects to be as much due to use of general best practice guidelines as the method of consultation used, recommending participation throughout policy development using a range of participatory methods, and a wide range of individuals.

Public Hearings

Public hearings usually take the form of a technical presentation followed by questions and comments (McComas, 2001). Although it might be possible to research the area in advance, it is likely the information presented will be new to some of the participants. Public hearings can be useful for gauging the level of support/opposition for a proposed scheme; provide a form of communicating information; a means of gaining insight and knowledge; and if the policy adopted is sensitive to the issues raised, the process may also increase the legitimacy of policy. The thinking of the participants may evolve during the hearings, but not everyone will speak and a number of participant views may be missed. There is also little structure to the questions and comments asked and these may not stick to the focus of the meeting. As such, comments are likely to suggest a range of feelings rather than anything representative. Indeed, it has been questioned whether those attending are representative of the population affected. Academic studies have shown that participants may have higher incomes (Sinclair, 1977; McComas, 2001) and be better educated (Sinclair, 1977) and also hold views different from those not attending and perhaps be more extreme in the expression of views (Gundry and Heberlein, 1984; Johnson *et al.*, 1993; McComas and Scherer, 1998; McComas, 2001). Where responses are extreme, heated exchanges may not be constructive to gaining understanding and may reduce trust in the process. In terms of outcomes, Chess and Purcell (1999) found public meetings to have frequently changed policy. Although perhaps positive, it is suggested by Chess and Purcell (1999) that this is due to the nature of public meetings in which participants are reacting to agency propositions rather than providing an input into policy development. Only in the case of one of the eleven studies they have considered did the public meeting yield a consensus.

Focus Groups/In-depth Meetings

Focus groups and less focused in-depth meetings are useful tools for considering the attitudes of the general public towards environmental issues (Burgess *et al.*, 1988a; 1988b; Burgess, 1996; Davies, 1999). Krueger (1994) defines a focus group as a: 'carefully planned discussion designed to obtain perceptions on a defined area of interest' (p. 6). Contrary to focus groups, in-depth group discussions do not have a protocol. Group members are given a broad theme and are asked to discuss the theme. They may meet a number of times, enabling participants to discuss the issues with friends and family as well as research the issue further. Unlike focus groups, the moderator does not drive the discussion. He or she only hands over the theme and allows discussion to flow at will. Both types of meetings provide a permissive, non-threatening environment that often enables frank discussion and as such these approaches have great learning potential. This can be useful for including groups who are often under-represented, for example, within public-hearings. Focus groups can also be useful as less tangible, moral and cultural environmental values can be considered. This was nicely demonstrated by Davies (1999) in terms of public consultation on environmental issues related to planning decisions, where a questionnaire survey had failed to 'reflect the depth and diversity of public's environmental valuation' (p. 304). By their nature, however, the information learnt is purely qualitative and these approaches should be used for understanding the range of values rather than as providing a representative indication of the importance of issues. In the same vein, they are also not designed for consensus forming in terms of policy decisions, with open outcomes. Trade-offs are also not usually considered within focus groups, such that it may be uncertain as to the strength of preference associated within the comments made, other than the tone of voice or expressions which can be somewhat subjective.

Citizens' Juries

As with focus and in-depth groups, citizens' juries also consist of a small group of individuals, in this case somewhere between 10–25 jurors. Unlike other group methods, where it is common to select groups based on similar experiences, citizens' juries require a 'representative'[1] group of citizens to be assembled (Coote and Lenaghan, 1997). Depending on the complexity of the topic area, the meetings usually take between two and five days during which the jurors can call 'witnesses' who present additional information to the panel of jurors and are able to cross-examine the witnesses. As such the jury is given a degree of power to define what information it requires. At the end of the meetings, the jurors will have learnt a great deal about the relevant issue and are expected to reach a conclusion on the matter considered, preferably based on consensus but if this is not forthcoming through the use of a voting procedure. Further to the decision on the particular scheme considered, qualitative information on the debate can also be analysed which both enables an understanding to be gained of the reasons for the decision(s) made, and also to determine the credibility of the process through assessing the extent to which the verdict was made freely or influenced by particularly forceful individuals. Citizens' juries have been used widely in Germany (where they are called 'planning cells') (Dienel, 1997) and the United States (citizens' juries is the United States term for these meetings) (Crosby *et al.*, 1986). Their popularity, however, has been increasing and they have been used in a number of countries, with many environmental applications (Brown *et al.*, 1995; Ward, 1999; Aldred and Jacobs, 2000). Citizens' juries are considered in detail in Chapter 8.

Questionnaire Surveys

A representative sample of general public attitudes can be achieved by undertaking a questionnaire survey. Strength of preference towards environmental issues can be considered using attitudinal scales (Kahneman *et al.*, 1999) and behavioural intention measured (Fishbein and Ajzen, 1975; Barr *et al.*, 2003). Keeping the same format across the sample provides uniform responses which can enable comparisons between population groups on environmental values held by the respondents.[2] Although such interviews will provide an indication of attitude, the respondents are likely to be ill-informed, elicited without enabling the respondent to discuss the issues with others and will be restricted to general principles or issues with which there is prior knowledge. Responses tend to be quantitative and restricted to structured response options, which may be limited in terms of the level of information they provide.

Stated Preference Methods

A further approach to consultation on environmental issues is using stated preference methods. This approach provides an extension to the questionnaire surveys described above in which respondents are informed of the environmental scenario considered (using words, pictures and sometimes even computer images) prior to being asked questions regarding their willingness to pay, often through taxation, for different policy outcomes. The output of stated preference approaches are valuations for the particular environmental goods and services considered and often a great deal of information regarding the meaning of these valuations. This is particularly the case when stated preference methods are combined with qualitative methods such as focus groups. The resultant valuations can be used within cost-benefit analysis to extend the process of estimation beyond that of the costs and benefits estimated from market responses. Using this approach, the efficiency of resource use can be considered. Stated preference methods have been widely used within the United Kingdom and the United States within policy formation in areas of agriculture, transport, flood alleviation, water supply, coastal defence, forestry and biodiversity (Hanley, 2001). Although gaining a cautious endorsement by a distinguished panel of social scientists assembled by the National Oceanic and Atmospheric Administration (NOAA) (Arrow *et al.*, 1993), the methods have remained controversial, receiving much criticism from economists, psychologists, sociologists and political scientists. This criticism has focused on issues such as the hypothetical nature of the valuations made, the cognitive load on the respondents and the perceived inappropriateness of the use of economic theory to consider scenarios of a communal nature. These issues will be debated in detail within this book. From the perspective of public participation, these methods can also be criticized as being reactive to policy rather than inclusive within policy development. However, as will be shown later in this book, the use of qualitative methods and experimental design can also be used to aid policy development. Furthermore, some degree of flexibility can be built into the method design to assess different policy options.

CONCLUSION

This short chapter has provided an introduction to the challenges of public consultation regarding environmental issues and an overview to the various approaches which can be used. The description has attempted to be illustrative rather than exhaustive in its coverage, giving an indication as to the key challenges of the methods described. One thing that is clear from the discussion is the difficulty involved in gaining a representative understanding of public opinion, when the issues are so complex and uncertain and respondents/participants need to be adequately briefed prior to stating their opinion. There are also many conflicts, in personal and citizen type values, but given the subjectivity involved consensus is likely to be difficult to achieve with many differences of opinions amongst the scientists, decision makers and general public.

NOTES

1. With such a small group this is likely to fall short of the requirements of opinion polls. Furthermore, given the length of time required to attend these meetings without any legal requirement to attend, this may further question the representativeness of the groups.
2. See O'Riordan (1985) for an example of how environmental value attitudes can be explored using questionnaires.

3. Stated preference: methods and challenges

INTRODUCTION

Economists have taken five principal approaches to non-market benefit estimation: production function; recreational demand models: hedonic pricing; contingent valuation; and choice experiment methods. Production function, recreational demand and hedonic pricing methods are all based on the link between expenditure on market goods and resultant implicit consumption of non-market goods and services, where respondents reveal their preferences for the non-market goods and services through market transactions. As they are based on observed behaviour they are ex-post techniques (revealing the policy outcomes), whereas, most management scenarios need to be considered ex-ante (prior to the implementation of policy). In the 'hopeless case' situation of strong separability, where changes in environmental quality or quantity have no affect on market good transactions these techniques are ineffective. Revealed preference methods can only be used for estimating use benefits.

The questionnaire-based methods of contingent valuation and choice experiments are based on expressed preferences elicited in constructed markets or referenda for the actual change in public goods and services considered. These methods can provide ex-ante use and non-use benefit estimates and are in theory highly flexible and applicable to a wide variety of goods and scenarios. However, their very flexibility invites application to complex problems intractable to other techniques. This in turn has highlighted a host of practical and theoretical problems reflected in a voluminous research literature. Indeed, stated preference methods have many critics, some of which have seriously questioned their application to non-market benefit estimation (Kahneman and Knetsch, 1992; Hausman, 1993; Diamond and Hausman, 1994; Boyle *et al.*, 1994 and Kahneman *et al.*, 1999). Despite these criticisms, cautious but positive assessments of their usefulness have also come from many sources (Mitchell and Carson, 1989; Arrow *et al.*, 1993; Carson and Mitchell, 1995; Cummings and Taylor, 1999). Given this controversy, there are many challenges faced when conducting stated preference surveys. In order that this debate can be understood, this chapter sets out the conceptual framework for estimating non-market benefits, introduces the stated preference methods of contingent valuation (CV) and choice experiments (CEs) and outlines the challenges faced in conducting such valuation studies.

PROVISION OF PUBLIC GOODS

From an economic perspective, the scenarios considered within stated preference studies often have the public good characteristics of non-rivalry and non-excludability. A good is non-rival or indivisible when one individual can consume a unit of the good without detracting from the consumption of others. An example of an indivisible good is a private beach or nature reserve, where up to some maximum use level, permitted access is enjoyed jointly. Non-excludable goods cannot be withheld from the enjoyment of others, which may include defence, carbon fixing and air quality.

In the case of public goods, economic theory suggests marginal cost relates not to the individual but instead to the total cost of supplying the good. In which case, for a given consumer, the marginal cost is likely to exceed the marginal benefits they receive and hence producers cannot recoup the costs of providing public goods. For these reasons, it was suggested by Samuelson (1954) that no decentralized pricing system could determine optimal levels of production and consumption for public goods. Public goods are consequently non-market in nature and are invariably provided through collective provision.

Ostrom (2000) suggests that experimental economics has been able to demonstrate the ineffectiveness of microeconomic theory in explaining behaviour concerning collective goods. Her review of experimental economics suggests a significant proportion of respondents following social norms when considering collective good situations. The social norms observed were reciprocity, trust and fairness, and are used to explain why collective action occurs. As stated preference is mostly applied to the valuation of public goods, with some form of collective payment, the expectation is that social norms will play a role in the responses elicited.

There is inevitably an information problem when providing public goods, in terms of what public goods to provide and how much. In the provision of such goods there is usually government intervention in the form of regulation and/or direct provision. Through cost-benefit analysis and the estimation of economic value some insight into how to answer these problems can be reached.

COST-BENEFIT ANALYSIS AND TOTAL ECONOMIC VALUE

Cost-benefit analysis (CBA) provides the present value of benefits minus costs, where they are measured in monetary terms (Pearce, 1983). Using CBA a project would be deemed economically efficient if it maximizes net benefits to society as a whole. Preferences are estimated in terms of ability to pay with typically the existing income distribution providing the implicit weighting for individual preferences. As such, no consideration is made of the equity of the resulting situation. Despite this and other problems, CBA can provide useful policy information (Hanley, 2001). For example, considering a project that would have a detrimental effect on the environment, if market costs are found to outweigh benefits then, without having to assess non-market amenity loss, the recommendation on economic efficiency grounds would be for the project not to go ahead.[1] In situations where market benefits outweigh costs, it is necessary to include some indication of non-market benefit loss, which requires the understanding of the concept of total economic value, which extends beyond market transactions.

Economic values depend on anthropocentric preferences and are estimated by considering what individuals are willing to forgo in the way of other resources in order to increase or maintain their well being or utility. Usually this 'opportunity cost' is expressed in terms of their willingness to pay (WTP). The total amount of resources that individuals would be willing to forgo represents the total economic value (TEV) of the improvement or preservation of the goods and services considered.

Total economic value can be categorized into use and non-use benefits. Use benefits are associated with the direct use of the goods or products. Recreation and education can also be obtained from direct use. Indirect use value derives from the services provided by the goods or natural environment considered, such as nutrient recycling, floodwater control or reduced crime from improved street lighting. However, for something to have an economic benefit it does not necessarily involve use. Indeed, Randall states that:

> an action has a *prima facie* economic benefit if an action increases the availability of something that is scarce at the margin and if that 'something' is desired by someone; that is, if it is at least potentially a source of human utility. (Randall, 1991: 303)

From this definition a range of non-use values can be determined. Although non-use benefits may be considered to be related to previous recreational visits or close residence, it is also possible to receive sizeable amenity benefits without actually visiting or living in the area. These benefits can be associated with the knowledge of the existence of the good and services from some other media, for instance books or television, or just purely in the belief that the goods and services considered should be preserved regardless of human use. Indeed, Krutilla (1967) suggests that individuals may: 'place a value on the mere existence of biological and/or geomorphological variety and its widespread distribution' (p. 781).

Hence this form of benefit is referred to as existence value. Although individuals may receive non-use value without actually visiting or using the goods and services in question, Larson (1993) suggests that it is unlikely that such benefits will be large in the absence of some behaviour on the part of the respondent. For example, a respondent with non-use value for an environmental good may spend time absorbing information from the media, change their consumer behaviour to reflect preferences[2] or even just think about the problem. Bequest value is also non-use and derives from the knowledge that the goods and services will be maintained for one's children, grandchildren or just future generations. Beyond these more conventional values, philanthropic value is associated with the satisfaction of ensuring resources are available to other members of the current generation.

Using the conventional TEV framework no new components of value emerge with the introduction of uncertainty. Nevertheless, in the presence of uncertainty, there may be a special case of 'quasi-option value'. Arrow and Fisher (1974) introduced this term and asked if the introduction of uncertainty in the costs and benefits of a proposed project would have any effect beyond the replacement of certain values with expected values. Arrow and Fisher (1974) suggested that: 'the expected benefits of an irreversible decision should be adjusted to reflect the loss of options it entails' (p. 319). Due to the irreversibility and uniqueness associated with some environmental goods, quasi-option value may be important.

In the case of environmental goods Turner *et al.* (2000) regard TEV as only a part of the 'total ecosystem value' (p. 13), where the overall system itself has a value beyond that of its individual components. Hence, for environmental goods, the economic values described in the framework above are dependent on the continued operation of the whole ecosystem. Hence, 'secondary' economic values are dependent on the 'primary' value of the entire system on which ecological functions and services depend. As such, choices of scenarios considered within environmental valuation should reflect an understanding of ecological sustainability. In theory at least, issues of social equity can be included within the TEV framework. For example, considering public support for disabled access to a recreation site, the utility of those benefiting from the scheme could be measured in their WTP. However, public support reflects a social norm or ethical view from 'society as a whole' that disabled access to recreational sites should be provided. As such, disabled access depends on the WTP of society to subsidize such schemes. In practice, however, such issues may be too plagued with ethical concerns of right and wrong and principles such that a sizable proportion of those interviewed may be reluctant to consider the situation in terms of trading-offs (Jacobs, 1997; Wilson and Howarth, 2002). Such 'social' issues will be returned to below.

ESTIMATING NON-MARKET BENEFITS

The stated preference methods of contingent valuation and choice experiments are based on expressed preferences elicited in constructed markets or a referendum for the actual change in public goods and services considered, and can be used to provide ex-ante use and non-use benefit estimates.

Contingent Valuation Method

The most popular expressed preference approach, contingent valuation (CV), involves constructing a hypothetical transaction to value the specific policy under consideration, where outcomes are contingent upon respondent choices (Mitchell and Carson, 1989). CV exercises concentrate on the valuation of a particular scenario that presents a potential quality change, environmental or otherwise. This requires researchers to concentrate on providing adequate information about the scenario for the respondent to be able to determine his or her preferences for a fixed quality change. The first CV wetland benefit study estimated use benefits from waterfowl hunting in the USA (Hammack and Brown, 1974). Since this pioneering study there has been a gradual widening of application, where both use and non-use values have been estimated. A key advantage of the CV method is that it measures directly the environmental change considered, which provides great potential for policy analysis. In terms of non-use benefits, the suggestion of a distinguished panel of social scientists assembled by the National Oceanic and Atmospheric Administration (NOAA) to produce a report to critically evaluate the validity of CV was that the method: 'can produce estimates reliable enough to be the starting point of a judicial process of damage assessment, including lost passive-use [non-use] values' (Arrow *et al.*, 1993: 42).

The theoretical underpinnings of this approach lie in utility maximization (Nicholson, 1989) and constrained optimization with respect to household income (y), and a vector of prices (p), such that if there is an improvement in a public good from z^0 to z^1 (that is $z^1 > z^0$) then household utility after the improvement ($\mathbf{u}^1$) is greater than utility before the improvement ($\mathbf{u}^0$):

$$\mathbf{u}^1 = \mathrm{v}(p, z^1, y) > \mathbf{u}^0 = \mathrm{v}(p, z^0, y)$$

The compensating variation (c) measure of the utility change is

$$\mathbf{u}^1 = v(p, z^1, y - c) = \mathbf{u}^0 = v(p, z^0, y)$$

It is the amount of money that can be taken away from the household after the improvement in the public good from z^0 to z^1 that will leave the household just as well off as before the change (Mitchell and Carson, 1989). This compensating variation can also be considered as the WTP for the change (Fisher, 1996).

WTP can be assessed using an open-ended approach that merely asks the respondent for the maximum amount they would pay in respect to the change in provision described. Although the open-ended alternative has advantages in terms of ease of analysis, information provided and the absence of distributional assumptions, it has been much criticized, for example, in terms of incentives for strategic behaviour, sensitivity to the 'fair-share heuristic'[3] and difficulty of the respondent task (Hoehn and Randall, 1987; Bohara *et al.*, 1998). To comply with the NOAA recommendations (Arrow *et al.* 1993), most CV practitioners in recent years have adopted a closed-ended format for the measurement of willingness to pay (WTP) for non-market goods. Using the closed-ended format, typically the respondent can choose between the 'with' policy situation at a given price or bid level (BL) and the 'without' at zero price. The yes/no responses to the BLs are modelled within a discrete choice framework from which welfare measures can be estimated (Hanemann and Kanninen, 1999).

The NOAA panel recommended the use of the closed-ended approach, where, in its simplest form, the respondent is faced with a dichotomous choice between the existing level of public goods with no increase in tax or alternative payment vehicle, and a specified improved level of public goods but with an increase in tax as the BL. If the household chooses that the scheme should go ahead at the tax amount stated then it is willing to pay at least that amount for the improved situation. Through making appropriate assumptions, different forms of welfare measure can be calculated from this information. Compensating variation is described above, but also the amount of tax that would result in 50 per cent/75 per cent of households supporting/rejecting the improvement programme may also be useful. This exposition can also be explained for a scheme to prevent or alleviate an action that would, for example, prevent environmental damage.[4]

Choice Experiments

Despite its potential, the CV method lacks the flexibility to accommodate for uncertainty and to value components of the scenario considered. Although, some flexibility can be introduced through varying the environmental quality considered by each respondent (Howe and Smith, 1994) and making sequential valuations, multi-attributed scenarios are perhaps better explored using an alternative methodology, such as choice experiments (CEs). CEs explicitly incorporate choices between attributes and attribute levels making the consideration of alternatives implicit within the valuation process.[5] As with CV methods, compensating variation can be estimated using CEs for both use and non-use benefits. Although CEs were only extended in the early 1990s to estimate the impacts on economic welfare from changing the provision of public goods (for example Viscusi *et al.,* 1991; Opaluch *et al.,* 1993; Adamowicz *et al.,* 1994), the flexibility of the methodology has led to an increase in its popularity (Bennett and Blamey, 2001).[6]

CEs are conducted in a variety of ways. One frequently used approach is the use of 'profiles' to describe the good or service being researched. Profiles report levels of the particular attributes used to describe the good being studied. Table 3.1 provides an example of such an approach with three profiles being presented (Alternatives A, B and C), where the attribute levels for the particular card presented have been chosen from an orthogonal set. The respondent has then either to choose the profile offering the greatest utility or to rank the alternatives in order of preference. If valuations are to be estimated, one of the attributes needs to be price.

Since individuals are asked to choose their most preferred profile from each set shown to them, in a similar way to the dichotomous choice or closed-ended CV approach above, a discrete choice random utility model (for example McFadden, 1973) is used to investigate how the choices relate to attribute levels. Central to the formulation of such models is the hypothesis that individuals make choices based on the attributes of the alternatives (an objective component) along with some degree of randomness (a random component). This random component is consistent with random individual preferences. It is also consistent with the realistic notion that the researcher only has a partial knowledge of the real structure of the respondent's preference, while the unknown component is assumed to behave stochastically. Respondents can also be asked to rank the profiles in terms of preference (Beggs *et al.*, 1981), however, this can be more challenging for the respondent and is controversial as it requires more restrictive modelling assumptions (Hausman and Ruud, 1987; Ben-Akiva *et al.*, 1991; Foster and Mourato, 2000). Also ranked responses have previously been found not to be consistent with economic axioms of preference (Foster and Mourato, 2002).

While this task is more complex than CV questions, it allows for increased flexibility in the analysis because the attributes themselves can be valued, as well as the overall scenario. Using the results of a CE, the attributes are valued individually with marginal valuations estimated for each attribute level. The challenge for the researcher is to choose attributes that comprehensively describe the key elements of the scenario, while at the same time ensuring that the experiment does not impose too high a cognitive burden on respondents.

ESSENTIAL ELEMENTS OF A HYPOTHETICAL TRANSACTION

Fischhoff and Furby (1988) characterize a transaction in terms of the good purchased, the payment made and the social context within which the transaction is conducted. In the case of public goods with non-excludability in delivery and communality of payment, the social context is implicit within the goods and services valued and the payment mechanism used. Social context is also implicit within the hypothetical elicitation mechanism through which respondents state their preference. Before considering these issues in turn, it is necessary to briefly take into account the general principles that apply when designing stated preference transactions.

Unlike other transactions, stated preference implies a hypothetical situation, where the link between respondent response to a stated preference question, policy formation and payment, may not feel as binding as other forms of transaction. As a general principle, it is crucial that the perceived linkages are sufficient for respondents to put in the required effort to consider their preferences, ability to pay and opportunity costs of payment. Furthermore, as the scenarios considered are complex and often novel to the respondents, it is also essential that the task be within their cognitive ability, considering also time constraints within a questionnaire survey. Lastly, given the collective nature of the goods valued, norms are likely to play a role in the responses elicited. However, the transactions should be designed such that they are as far as possible follow existing norms. Where the transaction goes against a norm, the social context rather than the specifics of the goods considered may dominate the decisions made. These general principles will be considered in more detail in Chapter 5.

Presentation of the Scenario Considered

In order for an economic transaction to accurately reflect respondent preferences it is important that the good purchased is well defined and understood by the consumer. When purchasing market products the nature of the good is usually spelt out by the packaging or through a sales assistant. In a political referendum information is usually provided on both sides of the argument through the media or other source. In a hypothetical stated preference transaction the interviewer provides this role, where the good definition or policy scenario is described in terms of the need, potential actions that could be taken and personal consequences in terms of the required financial outlay. The stated preference approach differs from either market or referendum transactions in that the role of the researcher is to provide the information in as neutral a manner as possible. An important issue in scenario presentation is whether the estimated measure ac-

Table 3.1 Example choice card

Attribute	Landscape and wildlife impact on woodland, fields and environmentally sensitive agricultural land due to reservoir construction or enlargement	Landscape and wildlife impact on wetlands due to changes in the level of abstraction.	Landscape and wildlife impact on rivers and streams due to changes in the level of abstraction	Level of service received by households Average likely occurrence of a hosepipe and sprinkler ban (lasting no more than 1 year) and is also an indicator of pressure and the possibility of supply interruption.	Change in what your household pays in annual water charges (not including wastewater / sewerage)	RANK
Alternative A	No change	No change	No change	1 in every 10 years	Extra £5 per year	
Alternative B	Moderate worsening	Minor worsening	No change	1 in every 10 years	No change	
Alternative C	Minor worsening	Moderate worsening	No change	1 in every 10 years	No change	

Note: Environmental changes are shown in terms of landscape and wildlife, where the impacts are labelled either no change, minor or moderate impacts as determined by experts. As an illustration, a minor change would lead to no more than a 5 per cent change in the number and range of species (fish, birds, other wildlife species and plants depending on the type of area) affecting no more than 50 hectares of land (1 football pitch is 0.6 hectare and 1 hectare equals 2.5 acres). A moderate change would lead to no more than a 10 per cent change in the number and range of species affecting no more than 100 hectares.

curately and fully corresponds to the 'specific domain of content', which is the structure of the transaction and the description of the amenity (Mitchell and Carson, 1989: p. 192). Normally testing for such content validity is subjective and assessed via peer review. As will be discussed in subsequent chapters, this can also be achieved through the use of qualitative methods.

Information defining the good and scenario considered entails a clear definition of the policy situation under investigation in terms of the: characteristics of the goods and services considered; quantity and quality levels in the status quo and alterative with policy scenario(s); scenario content and policy background; the time period over which the change(s) will occur; the precise location of the change(s); and the method of provision. Bearing in mind limitations in respondents' cognitive ability, it is the task of the researcher to make sure that each respondent is aware of these focal aspects of the scenario. Clearly this task is very challenging. Too little information can mean that the respondent has an inadequate understanding of the scenario considered, but care must be taken as too much information can confuse respondents (Ajzen, *et al.*, 1996).

Design requires the testing of the construal process undertaken by the respondents (Fischhoff *et al.*, 1999). If the respondents' default assumptions or predetermined opinions about a good or service are seen to be largely correct, then little information is required. Where respondents' default assumptions are wrong and identified as such, they can be corrected with the use of carefully chosen information. Also it is important that respondents interpret the information as intended by the researcher. For example, it is hoped that within a stated preference transaction respondents will accept the scenario as feasible and able to deliver the required level of environmental protection or change. However, Fischhoff *et al.* (1993) suggest that respondents may perceive uncertainties in terms of scheme outcomes; and Powe and Bateman (2004) finding such perceived uncertainties to be a key determinant of WTP and sensitivity of responses to specifics of the scenario considered (see also Chapter 6).

Payment Vehicle

Within any transaction the purchaser is required to make a sacrifice, usually monetary, in order that trade is achieved. In the stated preference context the method by which the sacrifice is made is referred to as the payment vehicle. Within real markets welfare measures may sometimes vary depending on the payment mechanism used; examples include credit card advanced booking or hire purchase agreements. As these payment mechanisms provide additional benefits in terms of payment by credit for the good purchased, this is consistent with economic theory. In the stated preference case, however, the payment vehicle used has been found to be a key factor determining WTP. In the case of early CV studies this difference was considered a problem with the method

(Rowe *et al.*, 1980; Brookshire *et al.*, 1981; Schulze *et al.*, 1981). Arrow (1986) and Kahneman (1986) disputed this finding, concluding that due to the public good nature of the goods and services valued using CV, further differences were to be expected than revealed when valuing private goods.

As noted above, following her review of experimental economics, Ostrom (2000) suggests a significant proportion of respondents follow social norms when considering collective good situations. This is also likely to be the case within expressed preference studies. In terms of the payment vehicle the norms of reciprocity, trust and fairness are likely to be relevant. As payment vehicles are frequently based on a compulsory tax system in which revenue is used for a variety of purposes, the link between payment and receipt of the good is often not clear. Although the compulsory nature of the tax ensures payment by all and reciprocity, the ambiguity between payment and delivery can reduce trust in the agency responsible for provision. Others may object to the implicit implications of the payment vehicle used. In the case of a progressive income tax, for example, a vote to pay £10 by a person on a low income would imply that someone with a higher income would pay a much larger sum. Where respondents do not agree with this value judgement, perhaps considering it to be unfair, they may be willing to pay less for the scenario considered. Hence payment vehicle choice needs to reflect shared norms of trust and fairness; otherwise, these issues may dominate the responses made.

Hypothetical Elicitation Mechanism

Having described the scenario considered and devised a payment mechanism, the constructed market or referendum like situation requires a set of rules governing behaviour and a means by which a transaction, hypothetical or otherwise, can be made. The main rule for a market transaction is 'if you pay you get the good' where the appropriate laws set by government ensure fair play. In the case of a political referendum the voter states their preference and rules ensure that necessary payments will be jointly made if some form of majority is achieved. In the stated preference context, the rules of payment are not always stated but are implicit within the transaction mechanism used. In the case of dichotomous choice or closed-ended CV method, which was recommended by the NOAA panel (Arrow, *et al.* 1993), the respondent can choose, in the case of an environmental improvement, between the 'with' situation at a given price or BL and the 'without' situation at zero price. This choice is clear to the respondent, where a positive response would increase the likelihood of the scheme going ahead.

The norm of fairness may also be relevant in terms of the pricing and services considered, the funds for which are likely to be raised through some form of collective payment. In the case of a collective payment, shared norms of fairness may imply that the price reflects the cost of the scheme and that the cost should be fairly split between members of the collective group. Understanding such norms prior to the survey may be crucial to minimizing protest responses.

CONCLUSION

This chapter has provided an introduction to environmental valuation and stated preference methods, outlining the type of benefits that can be estimated, the methods for their estimation and the essential elements of a hypothetical transaction. The difficulties involved in design suggest the importance of good pre-survey testing and understanding of the motivations for responses; incentives for respondents to give meaningful responses; ability of respondents to give meaningful responses; and the presence of social norms due to the collective nature of the payment vehicle. It will be argued in subsequent chapters that this can be achieved using a combination of qualitative and quantitative approaches. This understanding will enable an assessment of how survey design can be improved and/or how appropriate valuation methods are to the particular scenario or policy situation being considered.

NOTES

1. Turner *et al.* (1983) provide a real life example where this occurred in Broadland.
2. The example given by Larson (1993) was participation in tuna boycotts (p. 381).
3. Here respondents do not state their maximum WTP but instead what they see as their 'fair-share' of the costs of the scheme.
4. See Mitchell and Carson (1989) or Hanley and Spash (1993) for a more detailed exposition of this theory.
5. See Boxall *et al.* (1996), Adamowicz *et al.* (1998), Hanley *et al.* (1998) and Garrod and Willis (1999) for comparisons between CV and CE.
6. See Farber *et al.* (2002) for some discussion of the wider context within which stated preferences can be used for environmental valuation.

4. Qualitative methods

INTRODUCTION

If practitioners believe that respondents are sufficiently able to articulate their responses to the questions posed, then valuation problems can be solved through improved survey design. This chapter looks at how this can be achieved using qualitative methods, with the main emphasis of the discussion being on perhaps the most popular method, focus groups. Although discussing the methodology behind qualitative methods, this chapter is designed to provide a 'how to do it' approach to focus groups in the context of stated preferences surveys.

QUALITATIVE METHODS

The use of qualitative methods differs from that of quantitative. In the case of quantitative economic valuation, a measurement of economic gain is made, which is the maximum amount of money that would actually be paid for the public good if the appropriate market or political referendum existed (Mitchell and Carson, 1989). The validity of such methods can be judged on the extent to which they accurately measure the theoretical construct and whether the sample is representative of the population of interest.

Using qualitative methods the researcher is not measuring or using an instrument as a proxy for a theoretical construct. They have no such restrictions and the purpose is not measurement but to gain an insight to improve the valuation process and to better understand the meaning of the valuations elicited. The complexity of issues and difficulties within stated preference are such that understanding and insight are required to sufficiently improve design and assess the applicability of the resulting valuations. Such understanding and insight would be difficult to achieve using purely quantitative methods.

With this different purpose in mind qualitative methods do not infer to a population, but instead try to gain a detailed understanding of the issues under investigation. As such, the results provide an insight into the range of issues considered where qualitative data should ideally be collected to the point of saturation, or rather, to where new insights are no longer being achieved. Although the frequency of an issue that is discussed within a series of focus groups may be indicative of its importance, Morgan (1997) warns against disproportionate response to isolated incidents in group meetings, suggesting the vividness of the direct contact may overplay an issue in the researcher's mind. Instead, there is a need for corroboration of qualitative findings through further research. How representative the insights are is best considered within quantitative research, where Carson *et al.* (1992) provide an illustration of a situation where a possible explanation for bias identified within qualitative research, was not corroborated within quantitative research.

Although the purpose of qualitative research is different from economic valuation, it can still be robust in terms of the methods used. Robustness is achieved through the use of systematic procedures within data collection and analysis, and openness regarding their description. As an illustration, checks of understanding can be made within the meeting by asking for confirmation from the participants; team debriefings used to cross-check interpretations of what was said; and systematic steps used within analysis such that there is an evidence trail, if required, allowing future verification.

There are many methods for the collection and analysis of qualitative research. For example, some approaches are based on language and how language is employed; others seek to establish a coherent and inclusive account of a culture from the point of view of those being researched; and others are aimed at theory building (Bryman and Burgess, 1994). Many share principles and approaches; and distinguishing between them is difficult. Their use is also much wider than the analysis of group or individual interview responses.

Group Interviews

Perhaps the most popular group-based qualitative method is the focus group, which has also become popular within stated preference studies. Morgan (1997) defines focus groups widely, to include a number of group-based approaches, where the researcher's interest provides the focus and group interaction provides the data. Krueger (1994) suggests that such groups should be small in terms of the number of participants and carefully planned to obtain perceptions on the researcher's defined area of interest. An alternative less focused in-depth method has also been developed in which participants meet on more than one occasion (Burgess *et al.*, 1988a; 1988b).

Compared with individual interviews, the history of focus groups in academic research is relatively recent. Although group-based methods had been used prior to the Second World War (Bogardus, 1926), major developments were made during the war by Merton and co-workers (Merton and Kendal, 1946; Merton, 1987) to explore the persuasiveness of propaganda efforts and the effectiveness of training materials for US troops. Despite earlier adoption within marketing, it was not until the 1980s that group methods, in the form of focus groups, became widely used to conduct academic research. In 1987 the first book on the topic was published for marketers (Goldman and McDonald, 1987), followed by an academic text the following year by Morgan (first edition of Morgan, 1997). Throughout the 1990s the popularity of the approach continued to increase.

The first known use of focus group meetings to aid stated preference design was as early as 1984 (Desvousges *et al.*, 1984), with the use of such groups to improve questionnaire design now being standard practice (Arrow *et al.*, 1993; Loomis *et al.*, 1993; Boyle *et al.*, 1994; Chilton and Hutchinson, 1999). More recently the application of focus groups has been extended to improving our understanding of the meaning of valuations elicited (Blamey, 1998; Brouwer *et al.*, 1999; Powe *et al.*, 2004a; 2005). The alternative in-depth group approach has also been used by Clark *et al.* (2000).[1]

Directed by a moderator, focus group meetings provide a permissive, non-threatening environment that enables discussion. This provides a key advantage of the approach. The importance of group homogeneity is emphasized by Krueger (1994), where generating group discussion is likely to be easier if participants have similar experiences in terms of the issues considered and socio-economic characteristics. Care must be taken when generalizing the findings of group discussions to the sample as a whole. This is partly due to the numbers involved being generally small but also to the social processes that make participant comments dependent on the group context from which they were made. Krueger (1993, 1994) suggests cautious generalizations can be made but percentages should not be quoted.

Individual Interviews

Qualitative individual interviews were developed as an alternative to structured questionnaires, with closed-ended response options regarded as too restrictive and potentially biasing responses (Rice, 1931). This led to semi-structured or non-directive interviewing, in a similar way to group methods, but as the individual approach to interviews was already prevalent, acceptance by the academic community was smoother and more immediate. Like focus groups, the use of open-ended questioning within questionnaire design has become common practice within stated preference studies (Lazo *et al.*, 1992; Arrow *et al.*, 1993; Kaplowitz and Hoehn, 2001). Similarly, the individual interview approach has also been used to gain an understanding of the meaning of the valuations elicited. Such research has borrowed from cognitive survey design literature and used verbal protocol analysis (Schkade and Payne, 1994) – a psychological research method that encourages respondents to 'think aloud' whilst completing questionnaires. Further to collecting concurrent discussion during the completion of stated preference questions, a debriefing can also occur at the end of the questionnaire using open-ended questions, for example: 'How did you come up with your monetary amount in the previous question?'. Such retrospective responses, although supplementing the concurrent remarks during the interview, have been widely criticized as potentially providing an inaccurate perception of respondent motivations (see, for example, Schkade and Payne, 1994; Fischhoff *et al.*, 1999). Instead, they can reflect how people think, having had time to rationalize the situation. Such retrospective responses may express motivations that are thought to be socially acceptable and reflect further thought having subsequently acquired additional knowledge and had time for reflection. Fischhoff *et al.* (1999) argue it is more appropriate instead to perform manipulation checks, which tests how closely the respondent construal of the information provided is to that intended by the researcher. Here open-ended questions regarding the task details, elicit, for example, the extent to which respondents have grasped the magnitude of the goods and services considered, the payment vehicle used and belief in the information provided.

Choosing Between Group and Individual Qualitative Approaches

The choice between group and individual-based methods is a difficult one, with Kaplowitz and Hoehn (2001)[2] illustrating that they provide different types and frequencies of data. However, this is only one study and, more generally, the issue of whether focus groups and individual interviews provide similar information has not been widely researched (Morgan, 1997). A key perceived strength of the group approach is that through information sharing during the discussion the understanding of the participants involved is greater. However, there is evidence that questions whether individuals actually do exchange all the information they have, instead, discussion tends only to involve information that is shared between the participants (Janis, 1982; Levine and Moreland, 1995). De Jong and Schellens (1998) suggest group methods may focus more on general problems, whereas individual interviews focus more on detailed/specific issues. This limited research suggests, if understanding of issues relating to a topic is the focus of the study, a combination of both processes would generate the widest range of issues and aspects of the problem.

As stated preference considers issues related to potential financial payments, such expression may be regarded as confidential and perhaps best considered in an individual setting, where decisions are only shared with the interviewer. This problem can be alleviated within a group session where the participants self-complete the questionnaire prior to the discussion and are not obliged to divulge issues to the group of a 'private' nature. However, using a group-based approach it is very difficult to consider concurrent remarks and, when considering respondent motivations, the discussion is limited to retrospective discussion. As noted in the last sub-section, when motivations are reported retrospectively there are potentially problems with recall and also 'post-hoc rationalization' of the responses given. Although concurrent approaches are also not without criticism[3] the interview approach does have the advantage of eliciting both concurrent and retrospective responses. As such individual interviews can be viewed as better for considering respondent motivations.[4]

Group discussion, however, may generate insights that would not otherwise have been observed. Group approaches differ from interviewing methods in that they enable communal discussion, where group members influence each other by responding to ideas and comments that may not have been considered during the interview process. Furthermore, Morgan and Krueger (1993) suggest that focus groups give the security of being amongst others sharing similar experiences. Hence, participants may feel more able to discuss problems in a group rather than a one-to-one interview situation. Indeed, possibly due to the support of other participants, Svedsäter (2003) found people in focus groups more willing to criticize. Hence, group-based methods may lead to a more in-depth discussion, although perhaps more of social–communal issues and as such, are much better for considering the public acceptability of stated preference approaches.

PURPOSE, DESIGN, IMPLEMENTATION AND ANALYSIS

In order to undertake successful focus groups it is necessary to have a clear understanding of the purpose of the meetings; design and implement the meetings such that meaningful data are collected; allow participants to comment, explain and share experiences and attitudes; and perform careful and systematic analysis enabling insights to be gained relating to the topic of the meetings. This section considers each of these issues in turn.

Purpose of the Meetings

When deciding what the purpose of the meetings should be, there is a need to understand the nature of the problem being addressed. From this, researchers can determine what information is needed and how that information will be used. As part of this process the researcher must decide whether the focus group approach is appropriate. In terms of stated preference surveys, focus groups are useful in order to:

- *Gain an understanding.* Johnston *et al.* (1995)[5] describe how using experience, categorical and contrast questions, shared understanding in terms of perceptions, categories and language can be discovered and how they often differ from those used by experts. This may avoid biases due to

misunderstanding within survey questionnaires. This approach may be particularly appropriate where the topic is relatively new to the researchers.
- *Pilot test a questionnaire or part of a questionnaire.* Following the completion of a pilot questionnaire, the appropriateness of the design choices can be considered as well as identifying any misperceptions in the scenario definition and related concerns.
- *Determine the additional explanatory variables.* As will be explained below, there is a need for indicator data to be collected in the main questionnaire survey on a range of potential motivations for valuation responses. Focus groups can help consider the range of issues that may affect stated preference and help improve their wording within the questionnaire design.
- *Interpretation of quantitative results.* Following the completion of a quantitative survey, focus groups can be used to help explain any empirical regularities or anomalies encountered.[6]
- *Consider the public acceptability of the approach.* During focus groups the purpose and use of the stated preference results can be explained to participants. This enables participants to make a judgement as to whether their responses are sufficiently accurate to be used as an aid within environmental decision making, and whether the overall stated preference approach is considered to be appropriate, not only in terms of accuracy but also in terms of other issues such as fairness.

The appropriateness of using focus groups should be judged on the basis of the specifics of the study undertaken. For example, focus groups are not necessarily appropriate when there is a need to consider sensitive issues. In such cases the individual approach may be more appropriate. In addition, although the focus group approach is versatile, it may not be suitable for all purposes. For example, Krueger and Casey (2000) warn against the use of focus groups where a consensus is required, as this may put pressure on participants to conform rather than provide a permissive environment for discussion. If consensus is required, an alternative approach such as the Delphi method (Curtis, 2004) or citizens' jury (Brown *et al.*, 1995; Aldred and Jacobs, 2000) may be more appropriate.

Design

Group configuration and design

Morgan (1997) suggests most studies using focus groups:

- use homogeneous strangers;

- have six to ten participants; and
- consist of three to five groups per segment of population considered.

It should be noted however, that this represents common practice rather than a standard to which conformity is necessary. It has been suggested that the selection of participants for each group that are similar in terms of background, but not necessarily opinions and preferences, is likely to lead to an easier flow of discussion (Jourard, 1964; Morgan, 1997). However, should the researchers be interested in how opinions differ between population segments it would have been necessary to undertake a separate series of three to five focus groups for each type of participant (male/female, young/old, rich/poor). Given the inevitable cost restrictions within projects, clearly careful judgement is required. For example, Powe *et al.* (2005) recruited participants for each focus group from a range of ages and socio-economic segments, with the common attribute being that they were all water customers.

In order to enable a free discussion to occur, care is needed during recruitment. Expertise can be a problem, for example, as part of the collection of the data for Brouwer *et al.* (1999) difficulties were experienced in one of the groups where a participant had studied ecology and raised problems in terms of the ecology of the scenarios presented. Indeed, rightly or wrongly, he so forcefully stated his objections that this discouraged others from discussing their preferences towards the environment, for fear their preferences were based on poor understanding. Groups that are mixed in terms of authority or status can also be a real problem. Differences in status are less likely when strangers are recruited however, class/social barriers may still deter others from expressing their opinions. Strangers are also less likely to have prior agreement on issues and their use may enable a freer discussion. To illustrate this latter point, Krueger and Casey (2000) give the example of people travelling a long distance who more readily interact and disclose information with people they meet because they are unlikely to meet again. In practice, ensuring that all participants are strangers is far from straightforward, as people will often only agree to come to the focus groups if they are able to bring a friend. Groups within which some participants know each other can discourage others from expressing their opinion. If friends have similar views on the issue then this may also affect the results of the meetings.

Although six to ten people may represent the normal focus group size, the number of participants needs to be chosen on the basis of the requirements of the individual study. If the participants are unlikely to have experience on the topic, more people will be required to sustain discussion. For a topic that participants are very familiar with, a larger group may constrain the ability of group members to discuss their preferences and opinions. As Krueger and Casey (2000) suggest, the important thing is that the groups 'are small enough for everyone to have the opportunity to share insights and yet large enough to provide diversity of perceptions' (Krueger and Casey, 2000: p. 11). Larger groups have a tendency to fragment, making the task of the moderator more difficult and necessitating a more structured approach to the meetings. Fragmentation effectively means that the discussion cannot be deciphered from the recording and may render the results meaningless. It is common for participants to be told this at the start of the meeting, as effectively when fragmentation of the conversation occurs, the exercise is of little use.

General practice is to undertake three to five focus groups for each population segment. The number of groups reflects budget constraints, but also, where possible, the principle of saturation, which occurs when the range of understanding has reached such a level that undertaking more groups is unlikely to lead the researchers to learn much more about how people feel about the issues considered. When this saturation is achieved, the purpose of the focus groups has been fulfilled. For example, Powe *et al.* (2005) found saturation to be reached within the fifth and sixth groups, such that the moderator could predict the range of responses to be gained. Interestingly, a finding in the first group meeting, which potentially could have been of great importance in terms of understanding, was not repeated within subsequent group meetings. This illustration provides a further warning against generalizing from the results of only a limited number of group meetings.

Having decided on the type of group meeting, the number of participants per group and number groups, a strategy needs to be devised for recruitment. Morgan (1997) suggest that, given the numbers involved, it is unlikely that the sample size will represent the population of interest, even if random sampling is undertaken. As such, although the researcher should endeavour to recruit from a range of socio-economic characteristics, the researcher should also recognize there will be an element of bias within the findings. An alternative strategy would be to focus only on those population segments considered information rich (Krueger and Casey, 2000) and particularly knowledgeable about the subject. In terms of stated preference, where the aim is to understand general public preferences and opinions, information is spread fairly uniformly across the population of interest. One approach is to recruit participants who had expressed a willingness to attend a group meeting at the end of their valuation questionnaire. Although this strategy can work well, in the case of the Powe *et al.* (2004a) example, the approach was found to recruit people either interested in the issues considered or to have particular experiences with the water company. Instead, Powe *et al.* (2004a) eventually had to adopt the alternative strategy of on-street recruitment with a screening questionnaire. Participants could also be recruited by house-to-house visits or the use of telephone lists. Experience suggests that face-to-face recruitment methods are superior, as the participant often feels they would be letting the recruiter down should they not attend. An alternative approach would be to 'piggyback' on an existing group, for example, through a school or club. However, consideration should be given to what extent this may bias opinion through recruiting people from similar groups. The individuals involved are also likely to know each other, which may cause the difficulties discussed above.

When recruiting, participants will wish to know at least the topic of the discussion. There is a need to make the purpose of the meeting sound important, for example, through the sponsoring organization. However, care is needed that this only encourages participants with a narrow range of interest, opinions or experiences to attend. Incentives, monetary or otherwise, are common and this may help reduce self-selection bias, where only people interested in the topic attend. There is also a need to make the meeting not sound too onerous and generally make people feel comfortable about attending.

Venue and choice of moderator

The focus groups must provide a permissive, non-threatening environment that enables discussion and this must also be reflected in the choice of venue. Market research firms tend to have their own specialized venues with two-way mirrors and even video equipment, providing the opportunity for sponsors to observe the group meetings and to use video clips within the presentation of the findings. Such an approach is not necessary for the purposes considered here and the use of video may even be detrimental, with Morgan (1997) suggesting it may present a greater invasion into the privacy of the meeting. As long as the background noise is manageable so that a tape recording can be made, a pub, village hall, living room or other such venue would provide an equally suitable environment and may be seen as more conducive to a relaxed discussion.

Meetings need to be moderated. Although they are likely to be expensive, experienced moderators can be hired. Given the training required to get such a moderator up to speed with stated preference methods, it may be more practical to use someone from the research team and practice on a pilot group first. Should an experienced moderator be unfamiliar with the stated preference methods being used, it is important that a second moderator from the research team also attends the meetings in order to ensure key issues are not missed and sufficient probing is given to issues of concern. Alternatively, if expert knowledge is required on ecology, for example, to help answer respondent questions, a second person could provide this alternative information role. This approach differs from the normal situation where there is an assistant moderator who is primarily a listener and note taker. Should anyone be used who is not familiar with focus groups, it is essential that they are reminded of appropriate and inappropriate behaviour, such that they do not affect the permissive environment for discussion.

Deciding the questions to ask and the structure for the meetings

Before deciding what questions or topics to be considered, it is necessary to decide on the degree of structure to give to the meetings. Imposing structure enables greater cross-group consistency and makes it easier to compare group findings. However, an imposed structure may not provide sufficient flexibility for new issues to be raised by the participants. Indeed, Merton *et al.* (1990) argue that effective focus groups should cover a range of topics beyond that which the researchers already know, leaving the possibility that issues are raised within the meetings that were not anticipated. A compromise suggested by Morgan (1997) that allows the interplay between inductive and deductive approaches, is to 'funnel' the discussion, whereby the meeting starts open and then becomes more structured as it progresses. This ensures the discussion is not artificially compartmentalized by the meeting guide and allows a degree of freedom for original issues to be raised. It is also useful as it allows participants to discuss any issues they feel they need prior to focusing on the specifics of the scenarios considered. Having had their say, participants are more likely to be co-operative.

A possible structure for the meetings is as follows:

- *Opening:* introductory dialogue and then begin with an icebreaker such as going round the table. It is also important to emphasize what people have in common and state rules such as only one person to speak at a time.
- *Introduce the topic for discussion:* this can be in broad terms to gain an understanding of experiences, attitudes and preferences.
- *Key questions:* usually four or five questions that focus on the specifics of the scenario.
- *Ending questions:* 'all things considered', the final position of the participants on the critical areas of concern, which will clarify their opinions. This may be in response to a summary by the moderator.

Such a structure can form the basis for a meeting guide. This is essential for the moderator and should follow a natural progression across topics. However, the moderator needs to be allowed flexibility, as due to the variation in focus groups, the natural progression may vary between meetings. Experience suggests the meetings should not normally last more than 1½ hours, but it might be worthwhile suggesting the meeting may last slightly longer than this, just in case the meeting needs to overrun. The guide should provide an estimated time for each question, where most time is allowed for the key questions. Such time is difficult to anticipate in advance of the first meeting, but is likely to be related to the number of people in the group and the extent of their prior engagement with the issue(s) considered.

A variation of the above structure is the post-questionnaire approach, which has been refined over a number of studies (Brouwer *et al.*, 1999; Powe *et al.* 2004a; Powe *et al.* 2005; Willis *et al.*, 2005). In its most elaborate form, participants are first asked to complete the questionnaire. This is done in simulated interview conditions, where any questions from the participants are answered but only as they would be in a one-to-one interview, preferably where pre-determined answers have been devised. This process may take 15–30 minutes depending on the questionnaire. Having completed the questionnaire, the meeting begins following a similar structure to that provided above. Issues related to the public good under consideration are usually considered first. Subsequently, the experience of completing and reactions to the questionnaire are considered, where discussion of problems encountered and motivations for responses is encouraged, along with any further queries on any aspect of the questionnaire or study. At the end of the meetings, participants are given the opportunity to revisit the questionnaire and make any changes they feel necessary using a different coloured pen. As the focus group meetings allow the participants to deliberate and ask further questions regarding the issues, this tests the adequacy of the initial questionnaire responses to deliberation. In some cases this has also enabled issues to be considered that are too private to be raised by the participants within the group discussion. This use of pre-and post questionnaires has similarities with the approaches adopted by Morgan and Spanish (1985) and Sussmann *et al.* (1991), where the extent of attitude change within group meetings was also explored.

The questions asked by the moderator need to encourage conversation and have the characteristics that they are clear, one-dimensional, jargon free, open-ended (to imply short replies are not sufficient) and short. Merton *et al.* (1990) suggest that questions should encourage specificity, such that, where possible responses should relate to experiences rather than less meaningful generalities, attitudes or intentions.[7] This is based on psychological literature that suggests attitudes perform poorly as indicators or behaviour (Ajzen and Fishbein, 1977). As an important reason for undertaking qualitative research is to understand the motivations behind stated preference responses, this issue is crucial.

Some suggestions for introductory dialogue:

- I would like to stress that there are no right or wrong answers to the questions that we will be discussing. If there were, we would go to experts and they would tell us the answers. Instead we are only interested in your opinions.
- I would like you to discuss and comment on these issues, where you are completely free to agree or disagree with what I, or what others present here today.

- There are only a few ground rules. Perhaps the most important rule is that only one person should talk at a time, or I won't be able to hear what you say on the tape and your views will be missed.

Some sample questions that can be adapted to the specific scenario are:

Impressions
How did you find ____?
What do you think about ____?
How did you feel about ____?
Describe your experiences of ____?

Motivations for responses
What features of ____ did /didn't you like?
What do you like best/least about ____?
What prompted you to answer that way?
Describe what went through your mind.
What one factor was the most important?

Probing
Would you explain that further?
Would you give an example of what you mean?
Would you tell me more about that?
Is there anything else about ____ that might be relevant?
Please describe what you mean.
Does anyone see it differently?
Has anyone had a different experience?
Are there other points of view?

Useful phrases
That is something we are interested in hearing more about.
One of the things we are especially interested in is ____.
Thinking back to ____.
Let us return to that later, since we will be considering that soon.
All things considered ____.

If there is an important issue that does not get raised in the meetings, it is essential that a prompt is included within the guide and that the moderator should raise it. Clearly, it is always better if it is raised by the participants, but if not it will be unclear whether it was an important issue that has just somehow been overlooked, or just not considered.

Implementation

Some means of recording the discussion is required. There are now digital recorders on the market that enable the meeting to be stored on your computer. Care should be taken that they have sufficient memory on a high quality recording setting for the full length of the meetings. As digital recorders do not transcribe the meetings for you, more conventional tape recorders are more than adequate. The most important feature of any recorder, however, is the quality of the microphone and sound reproduction. It is recommended that two recording devices are used, because without an adequate recording a great deal of data will be lost. Prior to the meetings the recording devices should be tested. For example, some require the setting of the recording volume and if not set correctly this will cause many difficulties in transcription. Within the start of the meetings it is important that the moderator openly mentions that a recording will be made, state how the data collected will be used, give assurances of confidentiality and gain agreement for recording.

Refreshments are a good idea for the meeting. Often people arrive having just eaten, but having the provision shows hospitality and welcome to the participants. If you are on a low budget, tea/coffee and biscuits can be cheaply provided and usually at least someone will take you up on the offer. Alcohol in moderation is usually acceptable, but will depend on the venue.

How the room is set up will depend on the preferences of the modulator. Usually it will be round a table or in a circle of chairs, either is fine. Prior to the meeting, it is useful to engage with the participants as they arrive. This engagement will help put them, and possibly also the moderator, at ease. Usually they should be greeted at the door, welcomed and thanked for coming. If there are two people administrating the meetings this is made a lot easier as one can welcome people as they arrive, whilst the other engages in small talk not related to the topic with the others. Standard practice is to try and position the most talkative participants next to the moderator, so that the moderator can reach across them, should they become too talkative. Those less talkative should be sat opposite the moderator where they can be engaged by eye contact. However, in practice achieving this may not be so simple. If reluctant to tell people where to sit, the moderator can, instead, position or reposition him or herself.

The job of the moderator within the meeting is primarily concerned with directing the discussion; keeping the conversation flowing; taking a few notes; checking that the topics in the guide have been adequately covered; making comments and noting points to be returned to later. A good moderator will keep the conversation on track, make sure everyone has a chance to share his or her opinions and above all, be a good listener. The moderator must believe the participants have the wisdom required for the research project and give the impression to participants that they are being listened to and that their opinions are important. A demonstration that the moderator has listened, by returning to an earlier point made, can help achieve this. They should also remain non-judgemental, not communicating any approval or disapproval. Where this is difficult, others may correct the naivety of the participants for the moderator. Given the nature of focus groups, a good moderator should also encourage participants to share perceptions and points of view. A list of probing questions is presented in the previous sub-section that can be used. These ask for further information, explore differences of opinion and check meaning. The job of the moderator is not easy and involves multi-tasking: listening to the conversation; cross-checking the issues in the meeting guide; and probing on key issues such that important information is not missed. Learning the questions to be asked prior to the meeting can help reduce the need for multi-tasking and helps improve the flow of the meetings. Moderators develop an ability to know when to ask further questions and when to keep quiet and let a pause occur, and especially with eye contact and body language, to encourage further points.

Individual characteristics of participants can present problems for the moderator. As noted previously, differing levels of experience might make other participants less inclined to talk. But hopefully this problem can be addressed within screening prior to the meeting. If there is a dominant talker the conversation can be deflected away from them, for example, 'thank you, does anyone feel different' or 'that is one point of view, does anyone have another'. As previously mentioned, if the moderator is sat next to them this may give more control. Some people will be better at articulating their thoughts and these people should be encouraged, as their insights may prove to be more useful. A rambler, is different however, they often use a lot of words and take forever to get to the point. If this happens, the moderator should discontinue eye contact and look at their meeting guide or other participants. As soon as the rambler stops talking, the moderator should ask the next question looking at the other participants. Encouragement with eye contact and even a direct question may help shy participants. For example, 'I don't want to leave you out of the conversation what do you think'. Some people like to think carefully before speaking. It may be worth spotting them and asking their opinion on key topics prior to moving on.

Following the meeting, it is important that a debriefing occurs, where the moderator(s) and assistant cross-check reactions to what has been observed. It may be that the assistant moderator noticed some lack of agreement missed by the moderator and giving the assistant moderator the opportunity to ask final questions may enable this to be resolved. It is important to allow time to refresh after each meeting. Ideally, the transcripts would be written up prior to the next meeting, just in case amendments are required. If the focus groups are being used as a means of questionnaire design, this can be very much an iterative process and reflecting of the findings between meetings can be important.

Analysis and Reporting

More generally, the approach taken to the analysis of the focus groups will depend on the purpose of the study. However, even if the purpose is purely exploratory, given the dangers of casual analysis, certain procedures should still be followed. General rigorous procedures for analysing qualitative methods apply to the analysis of focus groups (Strauss and Corbin, 1998; Miles and Huberman, 1994; Bryman and Burgess, 1994; Coffey and Atkinson, 1996; Bryman, 2004). As noted previously they are broad in application and distinguishing between them is difficult. Below the general principles are discussed.

The author has tended to adopt the following procedures within his analysis:

- debriefings between the moderator and assistant/second moderator are held after each focus group meeting in order for first impressions to be considered;
- taped discussions are transcribed (best conducted by someone who was there), with some sections being carefully omitted if they have no relevance to the study;
- each comment is coded such that it is traceable to a specific individual in a specific group, for example, G1A refers to participant A in Group 1;
- themes are identified from notes and transcripts, where the text is sorted on a word processor (similar to the long table approach suggested by Krueger and Casey (2000));
- the word processed document is then used within the writing of the final analysis, where emphasis is based on a combination of theoretical and empirical issues from previous studies; the number of groups within which the issue was raised; how many people mentioned the issue; the energy and enthusiasm among the participants; and
- the self-stated importance of the issue.

The only computer package used by the author has been a word processor. This provides facilities to search for words and move sections of text. Word processors also provide sort facilities should they be required. Berg (1995) suggests that unlike quantitative methods, where the use of computer packages is essential, no equivalent computer packages have been developed that can understand the context within which words are used and interpret whether consistent meaning has been achieved. However, some dedicated packages are available and, as they provide clear directions as to how to analyse data they may prove useful. For example, textual analysis tools include ETHNO, NUD.IST and ATLAS/ti.

There is confusion in the literature about the level of analysis of group-based interviews. Morgan (1997) clarifies this by suggesting that neither the individual nor the group should be considered as a separable unit of analysis. What individuals say within a meeting is influenced by the group context, but also what happens in a group is influenced by the individuals attending that meeting. As such, the coding method used above, reflects both the individual and the group within which the comment was expressed. Whether presented in the final document or not, it is good practice to have an earlier draft that identifies the participants expressing a particular view. This provides a verifiable evidence trial which is available should it later be required.

Within the write-up, a careful balance is required between direct quotation from participants and a discussion of the issues. We have found sponsors who particularly like quotations because it gives the impression of closeness to the topic. However, overdoing quotations may make the report over-long and difficult to read. Decisions about quotes might be determined by their self-reported importance and/or the need to help clarify the meaning of issues raised.

TESTING UNDERSTANDING

Assessing the factors determining valuation responses can be important in terms of validity assessment, considering the extent to which the stated preference estimates relate to other measures as predicted by theory (Mitchell and Carson, 1989). This is commonly conducted using regression analysis and/or a comparison of mean WTP estimates under conditions which theory would suggest different values, with the latter being less common due to the cost of split-sample approaches. However, in the absence of qualitative methods, the choice of variables to include in the questionnaire to later test as potential determinants of WTP is based purely on expectations from theory, previous studies and introspection. This has the drawback that issues not included within the questionnaire or split-sample design cannot be tested as determinants of WTP within the quantitative models.

As discussed in the previous section, the use of focus groups within the questionnaire design process can allow flexibility for new issues to be raised, with the results of this more inductive approach further aiding choice of variables for survey questionnaires. This is particularly important when considering issues that have not been widely researched using stated preference methods. Qualitative methods can also be used post-survey, however, this does not allow for quantitative corroboration within the survey for issues raised.

Given that experiences are likely to better explain behaviour they should perhaps be given a priority within the analysis and relevant information collected wherever possible (Fishbein and Ajzen, 1975). However, findings from focus groups have illustrated that attitudes may also be important determinants of WTP. A recent innovation is to formally use attitudinal statements to cover a range of issues relevant to the explanation of WTP, with factor analysis being used to help determine the structure of the data and the orthogonal factors being included within the valuation models (McCelland, 2001; Nunes, 2002; Nunes and Schokkaert, 2003; Powe *et al.*, 2006). The factors motivating the valuation responses are derived from individuals' responses to attitudinal questions on a Likert scale. The Likert scale attitudinal statements are formulated with the assistance of focus groups, which provide an indication of the range of issues relevant and illustrations of respondent language. Consistent with the best practice from psychological research (Oppenheim, 1992), a series of attitudinal statements are formulated for each motivation factor. The statistical significance of the motivational factor coefficients can be used to assess their importance as determinants of WTP and the size of the coefficients determining the magnitude of the value.

CONCLUSION

This chapter has shown how qualitative methods can be used to improve the conventional stated preference valuation process. Qualitative methods were described in general before focusing on perhaps the most popular method, focus groups. Although the purpose of focus groups differs markedly from economic valuation, the two approaches can be seen to complement each other, with the qualitative research aiding understanding and help choose potential motivational factors for survey questionnaires; pilot test the questionnaires; and consider the public acceptability and appropriateness of the overall valuation approach. This can only be achieved, however, by following best practice. This chapter has provided an understanding of such best practice and, given the flexibility of the approach, it is left to the researcher to design the focus groups in a manner that both follows guidelines and is appropriate for the specific valuation tasks considered. There is no set approach, which can be pulled off the shelf and used in its entirety. Instead, focus group design requires careful consideration and adjustment.

NOTES

1. This approach differs to focus groups in that they are undertaken repeatedly with the same group. This allows for opinions to be formed and consideration to be given to the issues between meetings achieving more depth than with other approaches. However, the detailed nature of this approach means that it is unlikely to be repeated until saturation with different group cohorts and, as such, it is unclear whether findings using this approach are more widely held.
2. See also Kaplowitz (2000).
3. Baker and Robinson (2004) noted the limited scope for probing responses during the interview without distracting respondents from the task of answering the valuation question. Clearly not

everyone is able to consider the issues and simultaneously provide an eloquent discussion of how they are deciding on their preferences.

4. See, for example, Fischhoff *et al.* (1993) for a discussion of these issues.
5. See also Chilton and Hutchinson (1999) for an elaborate example of how focus groups can be used to identify respondent misperceptions, with this information being used within the questionnaire design process to help decide which information to include. Given the time pressure involved in survey questionnaires this is an excellent, though expensive, way of selecting the information to be included.
6. As an illustration of the usefulness of this approach, Morgan (1997) suggests he has a simple answer to any researcher struggling to understand why respondents have replied as they did in a structural questionnaire – 'why not ask them'.
7. Cummings *et al.* (1986) and Johnston *et al.* (1995) have also made a similar argument within the stated preference literature.

PART II

Analysing the results of mixing method studies

Introduction

Having introduced stated preference and qualitative methods in the last section, this section considers the results of mixed method studies. The section contains two chapters, the first of which provides a general analysis of the results from qualitative studies. The second gives specific consideration to the important validity issue of scope sensitivity. Through consideration of both chapters the reader is able to form an opinion regarding the applicability of stated preference methods to the valuation of environmental goods and services. Innovative use of group-based methods to address some of the difficulties raised within this part are considered within Part III.

Summarizing the chapters in more detail, Chapter 5 is the longest chapter in this book. Prior to considering the results of qualitative studies, the chapter provides a framework for considering difficulties encountered using stated preference methods. This framework is then used to systematically review the results of a wide range of qualitative studies which have used qualitative methods to better understand the meaning of valuation responses and general public reaction to the methods used. Within the context of the findings from qualitative studies, Chapter 5 provides an indication of the type of scenarios which are most/least likely to be applicable to stated preference research.

Chapter 6 considers the issue of scope sensitivity over which there has been much debate and is seen by critics to be a key concern regarding the validity of stated preference methods. Whilst opinions remain divided over whether stated preference research provides valuations sensitive or sufficiently sensitive to the characteristics of the goods and services considered, Chapter 6 does not debate this further but instead explores through the use of mixed methods the reasons why a proportion of respondents do seem to neglect the specifics of the scenarios considered.

5. Interpreting stated preference responses

INTRODUCTION

Having outlined the challenges of designing a stated preference transaction within Chapter 3, the discussion now turns to how stated preference methods can be interpreted. Chapter 4 has shown how to gain a better understanding of how respondents discuss and conceptualize the goods and services valued; and how qualitative methods can be used to help understand the meaning of stated preference responses by:

- exploring respondent motivations and strategies;
- considering attitudes towards the responses made; and
- considering reaction to the survey.[1]

This chapter considers these issues, providing a summary of the key findings within previous post-questionnaire qualitative studies. Some issues are shown to have cross-study importance and relevance for future valuation research, whilst others have also been found to be important, but only to specific types of application. More emphasis will be given to the former.

In terms of the structure of this chapter, the first section considers how to interpret stated preference responses in terms of economics, behavioural psychology and more social/institutional approaches. The second section then describes the methodology used to consider the qualitative findings from a number of previous studies. An analysis of qualitative findings is provided in the third and fourth sections, which are divided into issues specifically relating to transaction design and those more generally relating to respondent reaction to the methodologies used.

DEVELOPING A FRAMEWORK FOR INTERPRETATION

Reflecting upon the wide-ranging stated preference literature, there would appear to be three key issues for consideration: ability of respondents to give meaningful answers; implications of the hypothetical nature of the transactions

Table 5.1 Interpreting stated preference responses

	Issues for consideration		
	Cognitive ability of respondents	Hypothetical nature of the transaction	Communal nature of the scenarios considered
Problem	Struggling to understand the scenario, transaction and how they feel about it	Lack of incentive to take the price seriously and put in the required effort	Following of norms and values rather than considering the trade-offs required
Cause of the problem	Complexity and novelty of the scenarios considered and transactions used	Absence of response-policy link, real payment, low price per household	Valuing public goods with communal payment
Empirical outcomes	Use of multiple strategies, looking for cues in the survey instrument, over sensitivity to task and context factors and lack of consideration of the scenario specifics	Lack of consideration of ability to pay, opportunity costs of payment and scenario specifics	Disagreement with the implied trade-offs, confusion regarding the transactions used, dominance of norms and values and lack of consideration of the scenario specifics
Methods for testing	Verbal protocols, split-sample testing and regression analysis	Real payment tests, split-sample testing and regression analysis	Observing protests and protest attitudes, split-sample testing and regression analysis

used; and communal nature of the scenarios and payment vehicles used. These issues are separated in Table 5.1, but should be considered together when interpreting stated preference responses. Although providing to some extent a caricature of opinion and approach, these three interacting issues provide a useful framework for interpreting the results of qualitative studies.

The first issue for consideration outlined within Table 5.1 (Column 2) relates to the cognitive task faced by respondents. Here it is important to understand the implications of respondents struggling to understand the scenarios presented and how they feel about the scenarios. It is also important to remember the novelty of the elicitation mechanism within which the respondents have to state their preference. The second and third issues (Columns 3 and 4) relate to the consequences of the responses made. Economists have tended to focus on the hypothetical nature of stated preferences and the absence of a real payment. This is seen to reduce the motivation of respondents to put in the required effort and consider the financial implications of the choices made. In the third column the communal nature of the scenarios and payment vehicles used are considered. Here, Blamey (1998) suggests respondents not only consider the financial consequences of the outcomes but also compare the options to their held values (a belief that a specific mode of conduct or end state should occur (Rokeach, 1973 – referenced in Blamey)), personal norms (such as 'I should do my bit to help the environment') and any social norms that hold within the interview situation. Although such concerns are also relevant to goods and services for which there is a market, due to the communal nature of the scenarios and payment vehicles used, norms and values gain prominence within stated preference responses.

Cognitive Ability of the Respondents

From the behavioural psychology research perspective, the focus is very much on the difficulty of the task and the nature of subsequent responses, where the behaviour of respondents depends on the degree to which they have prior experience with the goods and services considered and the extent to which the stated preference exercise is a process of preference discovery or research aiding value construction. Drawing from behaviour psychology research, Payne *et al.* (1992) reports how respondents in a situation of value construction tend to have behavioural consistencies. For instance, a key issue is the sensitivity of responses to the task and context of the questions considered. It is argued by Payne *et al.* (1992) that good decisions occur where the choices made are consistent and are not susceptible (or invariant) to small changes in the way the questions are asked (task) or the options are presented (context). Tversky *et al.* (1988) suggest that invariance to minor changes in task and context will only occur when people have well articulated preferences and/or where people use a consistent algorithm or heuristic. Behavioural psychologists have often found constructed preferences to fail to have such invariance. Payne *et al.* (1992) suggest this is particularly the case when the task is complex in terms of conflicting values, the scenarios considered are complex, and where there is uncertainty regarding how the decision maker feels about the issues considered. Such task complexity leads to respondents struggling to deal with the problem and using 'relatively weak' heuristics (Payne *et al.*, 1992: 92), where heuristics are principles by which people reduce complex tasks to simpler judgemental operations (Tversky and Kahneman, 1974). Dealing with complexity inevitably leads to some information being neglected, some of which may be relevant to understanding the problem. For example, the availability heuristic occurs where it is assumed that the probability of occurrences is related to the ease with which such instances or information can be brought to mind (Tversky and Kahneman, 1974). Clearly, the 'ease of access' does not necessarily relate to the actual importance of this information in the decision being made. A second example occurs where respondents avoid conflict by using lexicographic reasoning that is cognitively simpler than explicit tradeoffs and easier to justify to oneself and others.

Relating these general behavioural findings to the stated preference research, the unfamiliarity of the scenarios and transactions used is likely to mean that value construction will be the norm, where Gregory *et al.* (1993) argue that the principal problem with stated preference methods is that they impose unrealistic cognitive demands upon respondents. Similarly, Vatn and Bromley (1994) argue that due to the complexity and uncertainty associated with environmental functions ('functional transparency'), it is not possible to present a clear definition of the actual resource valued. Schkade and Payne (1994) suggest that beyond the understanding of the scenario considered, respondents also have the difficult task of understanding how they feel about these often unfamiliar situations.

Whether, despite the lack of well-defined prior preferences, respondents can still give meaningful answers is of central concern. Stated preference practitioners generally acknowledge that prior to being interviewed respondents do not have well-defined preferences for some of the goods considered (Mitchell and Carson, 1989; Hanemann, 1994), but, rather than rejecting the process, they focus their efforts on improving questionnaire design and avoiding scenario misspecification such that stable and robust preferences can be constructed. Clearly, choosing between different perceived implications of preference construction is a matter of judgement, which is likely to depend on the available evidence. The evidence from behavioural psychology suggests that robust preferences are unlikely to be constructed.

Within the CV literature there are numerous examples of the invariance criteria not being upheld. For example, ordering/sequencing effects have been empirically demonstrated which may be larger than expected by economic theory (Kahneman and Knetsch, 1992; Powe and Bateman, 2003); preference reversals have been observed when comparing choice situations to valuations through CV (Irwin *et al.*, 1993); and open-ended responses tend to be systematically lower than those generated through the closed-ended CV approach (Kealy and Turner, 1993; Bateman *et al.*, 1995; Boyle *et al.*, 1996; Frykblom and Shogren, 2000). There has been little testing within CV studies of invariance to the framing of goods and services,[2] however WTP estimates have been found to be sensitive to the content of the description (Bergstrom *et al.*, 1989; Whitehead and Blomquist, 1991; Ajzen, *et al.*, 1996). Neglect of important information is related to this debate. Whether CV responses are sufficiently sensitive to the scope of the goods and services valued has been much researched.[3]

Although it is possible to explain at least a degree of invariance using economic theory, it does significantly complicate the use of the results within policy making. Interestingly, however, there is empirical evidence suggesting invariance to task being upheld when valuing private goods with which respondents have experience (Kealy and Turner, 1993; Boyle *et al.*, 1996). This may give some credence to the arguments of behavioural psychologists.

From a quantitative perspective there is a need to test for task and context sensitivity as well as insensitivity to the specifics of the scenario considered. These quantitative approaches (regression and split-sample testing) can be supplemented using qualitative methods by considering issues such as task complexity, perception of how accurate responses are, and strategies/heuristics used by the respondents to make the task manageable. In the case of the latter, whereas economists would consider the extent to which strategies/heuristics used are consistent with theory, behavioural psychologists would consider the extent to which they are consistent with focusing on cues within task and context to simplify the cognitive task, and the extent to which heuristics are used that do not require the consideration of scenario specifics. Qualitatively, behavioural psychologists have tended to use verbal protocols elicited within individual interviews (see Chapter 4). Focus groups provide more of a social setting for considering these issues and may be more relevant for considering the social aspects of the stated preference scenarios.

Hypothetical Nature of the Transaction

Further to expressing the need for significant effort when making decisions regarding public goods, Varian also expresses concerns relating to the hypothetical nature of stated preference, by stating: 'talk is cheap and economics is concerned with scarcity. Perhaps this is why economists have little faith in concepts like "stated willingness to pay"' (Varian, 1999: 241–2).

To economists the hypothetical nature of the choices made within stated preference are a key concern and as a general principle it is crucial that the perceived linkages between response and policy-formation are sufficient for respondents to take the price seriously and put in the required effort to give meaningful answers. The hypothetical nature of the stated preference transactions, where the link between response within the questionnaire and payment may not feel as binding as with other forms of transaction, is a key reason for respondents demonstrating insufficient effort. Study design efforts need to focus on providing a scenario that is incentive compatible providing truthful preference revelation (Hoehn and Randall, 1987; Carson *et al.*, 1999). In addition, combining the task complexity and the hypothetical nature of the scenarios, there is a need not only for respondents to take the amounts seriously, but also that the price is sufficient for respondents to put in the required effort.

Numerous studies have considered the extent to which stated preferences are similar to revealed preferences observed in real markets or referenda, with the findings mixed, with some studies suggesting a close correspondence with real markets and referenda (Johannesson *et al.*, 1997; Cummings and Taylor, 1999; Johnston, 2006). However, other studies have suggested a positive bias (Neill *et al.* 1994; Frykblom, 1997; Ajzen *et al.*, 2004; Willis and Powe, 1998). Alternatively, the quality and meaning of stated preference responses has been explored using regression analysis and/or a comparison of mean WTP estimates under different treatments that theory would suggest should lead to a difference in the WTP.[4] Given sufficient incentives, respondents should, for example, consider their ability to pay and the opportunity costs of payment in terms of consumer goods and services forgone and/or alternative public goods that could have been funded. It is also important that responses reflect respondent preferences for the specific characteristics of the goods considered. Failure to do so would question the use of subsequent valuations within cost-benefit analysis.

These approaches are post-survey and rely upon sufficient data being collected during the main survey to be able to test the various motivations. Hence, for example, the regression approach can be limited in terms of the range of explanatory data collected and the understanding provided by the significance of what can be simplistic proxy variables. Alternatively, split-sample comparisons are expensive to conduct and limited in terms of the range of issues considered. Split-sample comparisons also rely on pre-survey choice of information to be collected. Taking advantage of the coverage of the population given by quantitative analysis and the depth of pre-survey exploration using qualitative methods may provide useful insights into whether respondents are sufficiently motivated when responding to stated preference questions.

Communal Nature of the Scenarios Considered and Payment Vehicles Used

Ostrom (2000) suggests experimental economics has been able to demonstrate the ineffectiveness of microeconomic theory in explaining behaviour concerning collective goods. Her review of experimental economics suggests a plurality of values held, two of which are particularly relevant here. First, some actors follow self-interest and behave in a manner consistent with economic theory. Ostrom (2000) termed such individuals 'rational egoists'. However, to varying degrees a significant proportion of individuals (40–60 per cent in one-off and finitely repeated experiments) follow norms of trust and reciprocity (termed 'conditional co-operators'). For such individuals, when asked to participate in a communal project, such as those considered within stated preference studies, they are likely to have a predisposition to follow these norms and agree to the payment. Failing to follow such norms, may lead to feelings of, for example, guilt. Results also showed that where individuals felt they could trust other players, they were much more likely to behave in a cooperative manner. In the case of stated preference, trust or lack of trust in the organization responsible for receipt of payment and delivery may significantly effect cooperation within the scenarios considered.

More generally, it is argued by Blamey (1998) that ecological economics should not restrict itself to the neoclassical model of human behaviour. One alternative approach would be to use Ajzen's (1991) theory of planned behaviour and this approach has been previously applied to CV (Ajzen and Driver, 1992; Ajzen *et al.*, 2004). This approach has proved to be useful in explaining hypothetical bias, Blamey (1998) found an adapted and extended Schwartz's norm-activation model to be more useful in this context (Schwartz, 1977). This is a psychological model of altruistic behaviour and, as such, is appropriate for considering cooperative behaviour where there are non-use values. The model suggests that the activation of cooperative norms, such as 'I should do my bit for the environment', are most likely when the respondent has awareness of need, consequences and responsibility. Denial of need, effective action and/or responsibility for payment can be used to justify a non-positive response and alleviate the guilt of not complying with the norm of cooperation. Blamey (1998) extended this model to take account of the communal nature of the public goods and services being valued. As it is common for the provision of public goods to be coordinated through an organization, often government, where there is compulsory payment, issues of trust in the organization to deliver and belief in the information provided become more prominent. Attitudes towards the organization can be based on past experience where respondents may feel they have little control over the activities of such organizations. Furthermore, delivery and payment will have implications in terms of the fairness, upon which there may not be a consensus. To reflect these concerns Blamey (1998) added to awareness of need, consequences and responsibility a further category of acceptance of policy initiatives.

Although, within stated preference surveys, it is up to the individuals responding which issues influence their choices (Hanemann, 1994), the following of norms may lead to the use of inconsistent response strategies, making valuations highly sensitive to the particular norm followed and make it more likely that respondents will be influenced by the social situation of an interview. For example, in the situation of a dichotomous choice CV question respondents in favour of the scheme may have a natural predisposition to cooperate and agree to the payment in order to avoid the guilt of non-payment and non-cooperation. They may also have a self-image of being associated with green values and going against this will be difficult. Respondents following such norms may give little consideration to the specifics of the scenarios considered, especially if the payment is seen to be non-binding and hypothetical.

It was suggested in Chapter 3 that some respondents could consider stated preference situations to be similar to that of a charity. Such behaviour would be consistent with 'judgement by prototype' behaviour, where how the choice situation is resolved is determined by the relevant properties of the situation (Kahneman *et al.*, 1999). This is a development from the use of the representative heuristic whereby the likelihood of an event is perceived to depend on the extent to which it corresponds with an appropriate mental model. Similarly, Vatn (2004) suggests classification or typification of a problem implies different situations and contexts that support different types of rationality. In addition, Vatn (2004) suggests that where deontic relationships exist, different approaches to reasoning would be used than those usually associated with the purchase of commodities in a market setting. For example, the social nature of environmental issues may encourage respondents to behave more as citizens than consumers,[5] doing what is right and proper rather than purely following self-interest. Alternatively, given the social nature of the issues considered, respondents could try to reveal their aspirations about themselves, rather than focusing on the specifics of the scenario considered (Vatn, 2005). This would indicate that a more social approach to value construction would be needed within the consideration of communal scenarios.

The use of the economic approach assumes that the respondents are able to trade-off financial gain or loss with environmental change, reducing the comparison of goods to a single metric of money. For example, if there are ethical principles or feelings of moral obligation involved in the scenario considered, some individuals may object to being asked to trade-off these principles against money. An early study by Rowe *et al.* (1980) found evidence to support the view that respondents rejected the notion of WTA compensation for environmental loss.[6] In terms of economics, this situation can be characterized as people holding lexicographic preferences, whereby respondents give absolute priority to one good and holding such preferences violates the axiom of continuity. Such preferences are acknowledged within the economic literature but regarded as rare.[7] However, within stated preference studies evidence of such behaviour within the more conventional WTP studies has been observed by

Stevens *et al.* (1991), Spash and Hanley (1995) and Lockwood (1998). These responses could alternatively be interpreted as protests against the valuation of the environment (as if it was a commodity) and the use of such valuation within decision-making.

Using psychometric scales on environmental attitudes, Spash (2000) explored these issues further, finding that respondents with positive WTP also held ethical views on the scenarios considered. Indeed, those with ethical views stated higher WTP than others. These responses were seen to reflect the situation where, in the absence of an alternative institution that respects such ethics/principles, showing support through the stated preference questions is the only way to show their 'token' support for the schemes considered (Spash, 2000: p. 202). As such, those with lexicographic preferences may have stated the maximum they could pay maintaining a minimum standard of living (however defined). As the amounts stated were not that large, Spash (2000) raises the concern that such responses would be misinterpreted as more conventional economic preferences. The reason for the use of the lexicographic approach could be because it is cognitively simpler than explicit tradeoffs and easier to justify to oneself and others. This explanation is more likely to occur when the financial commitment is not viewed to be real.

METHODOLOGY

The analysis is conducted below using the following stages:

- formulation of aims;
- choice of studies; and
- analysis of findings.

Aims

The aim of the analysis in the next section is to explore the meaning of stated responses given, in terms of the task complexity, motivations and strategies for responses given and reaction to specific aspects of the transactions. The aim of the penultimate section extends this analysis by considering the general reaction of the respondents to the surveys undertaken, in terms of whether they thought the methodologies were 'fit for purpose'. This has been a more recent extension to qualitative analysis, but is very relevant when considering the applicability of the methods.

Table 5.2 *A list of quantitative studies used within this chapter*

Study	Benefits	Scenario	Scheme considered	Elicitation method	Payment vehicle	Qualitative method	Qualitative analysis
Schkade and Payne (1994); Schkade and Payne (1993)	Non-use	Migratory waterfowl	Covering of waste oil holding ponds	Open ended	Product prices	Interview (concurrent and retrospective)	Verbal protocol
Fischhoff *et al.* (1993); Fischhoff *et al.* (1999)	Use and non-use	River water quality	Clean up scheme	Open-ended	Product prices	Interview (manipulation checks)	Not stated
Vadnjal and O'Connor (1994)	Use and non-use	Undeveloped volcanic island	Avoid development	Open-ended	Trust fund	Interview	Content
Cameron (1997)	Use	River water quality	Reducing the flow of nutrients	Open-ended	One-off tax surcharge	Focus group	Not stated
Burgess *et al.* (1998); Clark *et al.* (2000), Burgess *et al.* (2000)	Use and non-use	Freshwater marshland	Land management agreements	Bidding game	National taxation	In-depth meetings and focus group	Grounded theory
Blamey (1998); Blamey *et al.* (1999)	Use and non-use	Freshwater marshland	Pipe to divert from drainage system	Dichotomous choice	Multiple	Focus groups	Content

Brouwer *et al.* (1999); Powe (2000); Powe and Bateman (2003; 2004); Bateman *et al.* (2001)	Use and non-use	Freshwater marshland	Saline flood alleviation	Dichotomous choice	National taxation	Focus groups	Content
Fischhoff *et al.* (1999); Welch and Fischhoff (2001)	Use and non-use	River water quality	Acid mine draining	Open-ended	Higher taxes and prices	Group meetings	Content analysis and verbal protocol
Svedsäter (2003)	Use and non-use	Global warming	Increase the cost of energy	Open-ended	Energy cost	Focus group and interview (concurrent and retrospective)	Content analysis and verbal protocol
Powe *et al.* (2004a)	Use and non-use	Biodiversity	Conservation and enhancement	Payment card	Water charges	Focus groups	Content
Powe *et al.* (2005); Powe *et al.* (2004b)	Use and non-use	Environmental attributes	Water supply options	Choice experiments	Water charges	Focus groups	Content
Willis *et al.* (2005); Powe *et al.* (2006)	Use	Safety from crime and road accidents	Street lighting improvement	Dichotomous choice	Local taxation	Focus groups	Content

Choice of Studies

Using database and Internet search engines, manual search through key publications and word of mouth, a list was assembled of studies using qualitative methods to consider issues relating to environmental valuation. The focus of the research and the ways that qualitative methods have been used varies between studies. Table 5.2 provides a list of the main studies upon which the subsequent analysis draws. Each of the studies provides some insight into the meaning of stated preference responses. In some cases an individual qualitative study has contributed to more than one paper and, for this reason, they have been reported together within Table 5.2. The table also provides a brief outline of the benefits elicited, scenario and specific schemes considered, details of the transaction design and the form of the qualitative analysis undertaken. A mixture of group and interview-based studies have been undertaken where in some cases both approaches have been combined (see Chapter 3 for a discussion of qualitative approaches).

Analysis

Using Fischhoff and Furby's (1988) transaction framework as a guide, the categories of discussion have been carefully chosen for the next section, where the analysis is divided into issues relating to presentation, payment vehicle and elicitation mechanism. In the penultimate section the research is divided into general reactions to the valuation process and assessment of the suitability of the values for use within policy making. The choice of subunits for analysis was based on whether an issue was raised within a study and how frequently it was raised. As individual responses are independent, using verbal protocol analysis the frequency with which an issue is raised is comparatively straightforward to assess. As there is no such independence using group-based approaches, the frequency that an issue is raised can be measured in terms of the number of groups within which an issue is discussed as well as the frequency of occurrence within the discussion. In order to be included, a unit of analysis needed to be relevant to the objective of the study, raised in at least two of the studies considered and frequently raised within at least one.

The analysis of the findings is provided in the next two sections. Following the mixed method philosophy of the book, if there are quantitative studies to support the qualitative findings presented, this is noted.

PROBLEMS OF TRANSACTION DESIGN

This section provides a summary of the key qualitative findings relating to transaction design and is divided into three sub-sections: presentation, payment and transaction mechanism. Throughout, the analysis will relate to the framework outlined in Table 5.1.

Presentation

As noted in Chapter 2, a key responsibility of the researcher within a stated preference survey is to provide sufficient information for the respondents to be able to state their preference meaningfully, where the presentation is understandable, and in as neutral a manner as possible. Following the framework of Table 5.1, as the goods and services considered are often complex and respondents have little experience at least in terms of stating their economic value, construal of the information provided within stated preference surveys is difficult. These difficulties may be compounded by the hypothetical nature of the transactions not giving sufficient incentive for respondents to put in the required effort and the complex communal nature of the payment vehicles used. In the light of these difficulties, Fischhoff *et al.* (1999) suggest respondents may misunderstand or embellish the information provided, where impact of this misconstrual in terms of WTP could depend on the disposition of respondents towards trusting or not trusting the information provided.

Sufficiency of the Information Provided

Given the constraints of stated preference questionnaires and the complexity of the scenarios considered, simplification within presentation is inevitable. As respondents differ in terms of their previous knowledge and experience, the shortening of the presentation to a manageable length is likely to cause some respondents to feel they lack sufficient information to state their preference. Considering the summary of the evidence provided in Table 5.3, it is clear that although in most studies some problems with information were identified, on the whole respondents appeared to be happy with the general information provided. Certainly this was the case in the most recent studies where the respondents were asked (Powe, 2000, Svedsäter, 2003; Powe *et al.*, 2004a; 2005; 2006) and in other studies it was implied from the replies (Schkade and Payne, 1994; Vadnjal and O'Connor, 1994, Blamey, 1998). For example, Schkade and Payne (1994) found only 5 per cent of the respondents and Svedsäter (2003) only 10 per cent to request more general information. In addition, Powe *et al.* (2004a) state that in three out of the four groups the level of information was appropriate for the task and, likewise, Powe (2000) reports a majority in five

Table 5.3 Comparison of qualitative studies (presentation, payment and delivery)

Study	Sufficiency of information	Misconstual of information	Denial of responsibility for payment	Trust in the agency responsible
Schkade and Payne (1994); Schkade and Payne (1993)	5% requested more general information; 14% mentioned absence of scheme cost (p=0.00)	None stated	Oil companies should pay (12%)	We will have to pay anyway (13%), Anti-oil company (7%) anti-government (3%)
Fischhoff *et al.* (1993); Fischhoff *et al.* (1999)	14% requested cost information; 7% more facts or greater clarity; 5% how would affect personally; 4% examples; 7% what caused the problem; 3% method of delivery; 1% how effective	21–31% did not remember scope of good; 13–46% problem could be eliminated; 77–64% alternative payment vehicle; most assumed information not presented	Not tested	Not tested
Vadnjal and O'Connor (1994)	Familiar with goods and services valued	None stated	Rejection of the scenario	Lack of trust government to deliver promises
Cameron (1997)	Difficulties of clear presentation	None stated	Polluters should pay	Widely stated; alleged previous misuse of funds
Burgess *et al.* (1998); Clark *et al.* (2000); Burgess *et al.* (2000)	Problems understanding good; Two respondents asked for cost information	Local residents perceived the payment vehicle to be a 'local tax'	Only a single respondent	A number of trust issues raised

Blamey (1998); Blamey *et al.* (1999)	Some requested good information; Requests for cost information within every group meeting.	Number questioned the realism of the scheme and other consequences within delivery	Suggested government responsibility or need to reallocate funds; asked if national/local responsibility	Some lacked trust to earmark funds for the scheme and often stated past experience
Brouwer et al. (1999); Powe (2000); Powe and Bateman (2003, 2004); Bateman *et al.* (2001)	Two of seven groups requested more good information; Requests for cost information within every group meeting	Questioning of the realism of the 'whole' scheme; confusion with payment vehicle within three groups	Discussed in every group, concern would protect farming and commercial interests rather than environment	Discussed in four groups; majority assumed money would be spent on scheme when replying (p>0.10)
Svedsäter (2003)	10% need more information; 7% cost information	14% misunderstood question; 10% uncertain of delivery	65% who is responsible and should pay	45% questioned whether could trust government to deliver
Powe *et al.* (2004a)	Problems understanding biodiversity; positive on the level of information	One participant misconstrued biodiversity	One group suggested a shared responsibility	Discussed in three groups; trust related amount of feedback.
Powe *et al.* (2005); Powe *et al.* (2004b)	Discussed in three groups	None stated	Discussed in one group	Discussed in one group
Willis *et al.* (2005); Powe *et al.* (2006)	Familiar good (street lighting); cost info. within presentation	One participant	Pay enough council tax already (three groups), urban residents benefiting more	Discussed in all four groups, previous wasted funds

out of seven groups to be happy with the level of information provided. Clark *et al.* (2000) provide an exception where the results suggest more information on the scenario was generally required. Similarly Fischhoff *et al.* (1993) reported that only 99 of the 212 respondents wanted no additional information. In addition, although the magnitude of the problem is unclear from his study, Cameron (1997) also reports participant concerns that the environment good can be sufficiently described for use within policy making.

Turning to the type of information requested, the second column in Table 5.3 provides some indication of the range of issues raised. Going into more detail than reported in Table 5.3, Blamey (1998) noted that some participants requested information on the quality of the environment protected, whether alternatives had been considered, whether there had been an environmental study and the environmental impacts of the proposed solution. The implication here is that some discussants felt that they did not have enough information at hand to give a meaningful answer. In two of the seven groups described by Powe (2000), a majority said that the CV survey should have specified in more detail how the flood alleviation problem could be dealt with, who would be affected and who would pay. In terms of choice experiments, Powe *et al.* (2005) report difficulties noted by a few participants suggesting there was not enough information to choose between the environmental attributes.

The above concerns do question the validity of the tradeoffs being made for some participants, but with the inevitability of having difficulties within presentation it is a question of reaching an acceptable level rather than eradicating the information problem. One solution would be to look beyond the limitations of the individual structured interview and consider a group-based approach. Indeed, by the end of the post-survey group meetings it was clear that respondents had gained a better understanding of the issues (Powe, 2000; Powe *et al.*, 2005; 2006). This is important when considering how to improve elicitation methods and the relative merits of group methods compared with individual interviews are studied in detail in Chapter 7.

Exploring the information problem further, the magnitude would appear to depend on the type of presentation method. Burgess *et al.* (2000) used a brochure with photos and text that was given to the respondents to look through in their own time prior to interview. Respondents noted that they felt they were under pressure to absorb this information quickly, not because of any poor interview technique but rather because the social situation was such that they did not want to waste the interviewer's time. Similarly, the results presented by Fischhoff *et al.* (1993) can be explained by the presentation method. Fischhoff *et al.* (1993) used a telephone survey that is inferior to the face-to-face approach as no visual information can be provided. The magnitude of the information problem may also depend on previous engagement with the issue. Studies by Vadnjal and O'Connor (1994) and Powe *et al.* (2006) in which engagement with the good was high, found few problems with the information presented.

Interpreting the meaning of these results in terms of behavioural psychology will depend on the extent to which the preferences are constructed during the interview. The results give clear evidence of constructed preferences (Clark *et al.*, 2000; Powe, 2000; 2004; Schkade and Payne, 1994). Again, the qualitative evidence suggests the problem of value construction to be greatest where respondents have little experience/engagement with the scenario considered. For example, Powe (2000) considered this issue for people living in close proximity to a wetland area under threat from saline intrusion, he found difficulties in constructing preferences to be greatest for those with little experience of recreation in the area. Given that the wetland area considered is of national importance and the scheme was to be paid for through national taxation, this demonstrates the problems for assessing the national value for the protection of the area. Hence, for studies where there was little prior engagement with the goods and services valued, preference construction will be the norm and the respondents' interpretation of what this means in practice will depend on the attitude regarding the ability of respondents to construct meaningful and stable preferences during the interview (Fischhoff, 1991).[8]

Although the general scenario information presented was often considered by respondents to be sufficient, lack of information on the cost of the schemes was commonly noted. For example, Schkade and Payne (1994) report 14 per cent of respondents to have mentioned the cost of the scheme was absent from the questionnaire. Fischhoff *et al.* (1993), Blamey (1998), Powe (2000), Clark *et al.* (2000) and Svedsäter (2003) also found further information on cost to be a common request. Cost information is not usually provided within stated preference studies due to its separate consideration within cost-benefit analysis. Hence providing cost information may lead to double counting. Using dichotomous choice CV Blamey (1998) and Powe (2000) both reported requests for cost information within every group meeting. The following comments noted by Powe (2000) illustrate the frustration of the participants. In one group a participant stated: 'I feel that there should have been some guidance as to what sort of money you were taking about for this scheme because I would have no idea what it would cost to repair a river bank'; and another; 'We are talking in a vacuum here because we don't know whether it would be 20–30 million or a billion'. By dividing the total cost by the number of people paying, respondents could then decide whether the bid levels offered are reasonable in terms of their 'fair share' of the payment. Such behaviour would reduce the meaning of the responses given and be inconsistent with the objective of eliciting maximum WTP. This issue is returned to later in this chapter.

Use and misconstrual of the information provided

The evidence provided so far has been based on respondent perceptions of their requirements. Further to satisfying respondent perceived requirements it is necessary that the process ensures a consistent understanding of the pertinent information relating to the scenario and transaction considered. As only a limited amount of information can be provided to respondents, they may not have sufficiently grasped the issues and, to compensate for this, may have filled in the gaps in their understanding with their own interpretation of the scenarios. For example, Welch and Fischhoff (2001) considered the use of social context information within a stated preference study and found a general insensitivity to the use of this information. However, the issue of information use has not been widely considered, and suggests an important area for future research.

Related to the issue of information use is its misconstrual. This issue has rarely been tested and may be something that participants find embarrassing to admit to. Fischhoff *et al.* (1999) suggest three possibilities where misconstrual of information can occur: a misunderstanding of the original presentation; an alternative perception of the information presented; and when respondents understand and accept the information presented but forget it when asked the stated preference question. In all three cases the respondents are answering a different question to that posed by the researcher. This will add to the standard error of the valuation estimates and if occurring systematically may cause a bias.

Exploring misconstruals using retrospective qualitative methods may give an inaccurate picture because individuals may have correctly understood and accepted the information when presented and during the stated preference question, but forgotten it by the time they are asked retrospectively.[9] Although this would recommend the use of concurrent verbal protocols as the best method for exploring misconstruals, identification of retrospective misconstruals may also be indicative of problems with the valuation process. Indeed, Fischhoff *et al.* (1999) suggest such a short-term memory of key issues would imply the processing of the information to have been fairly shallow.

Using manipulation checks (see Chapter 4 for a description), Fischhoff *et al.* (1993) considered misunderstanding of the original presentation of a scheme to improve water quality. Perhaps alarmingly, Fischhoff *et al.* (1993) found between 21 and 39 per cent (depending on the treatment) of respondents to incorrectly recollect the approximate number of miles of river that would be improved by the scheme. This was despite a reminder. Interestingly for the issue of scope sensitivity (considered in detail in Chapter 6), those respondents who thought they were getting more miles of improved river quality were also willing to pay more. To the author's knowledge this has not been tested in any other study.

Another misconstrual issue of relevance to the scope of the good valued is the perceived likelihood of the scheme successfully being delivered. Despite the presentation suggesting certainty of scheme delivery, Fischhoff *et al.* (1993) and Powe (2000) found this issue to be important. In fact, both Fischhoff *et al.* (1993) and Powe (2000) found respondents to be more optimistic about the success of more modest schemes, with those considering the schemes to be realistic willing to pay more. Within the extended analysis of Powe and Bateman (2004), scope sensitivity was found to occur only for those considering the schemes to be realistic. The perceived divergences in presentation and perceived scheme realism were also observed in other studies not specifically considering the issue. For example, Blamey (1998) reported that respondents questioned the effectiveness of the scheme to protect the environment and Burgess *et al.* (2000) report that due to insufficient information within the presentation, respondents were left with feelings of uncertainty as to how effective the scheme would be. Scheme effectiveness was also questioned by Powe *et al.* (2006). Hence, the perceived realism of the schemes is important when trying to understand the meaning of values elicited as well as when testing scope sensitivity.

A further misconstrual relates to the payment vehicle used. This occurred in both Burgess *et al.* (2000) and Powe (2000), where a national payment vehicle was being used to consider local schemes of national environmental significance. Although the payment vehicle used was consistent with reality and, in the case of Powe (2000) at least, was the preferred means of payment by most of the respondents, as the cost per household in the United Kingdom would be minimal, there was confusion as to what the actual payment vehicle would be. Finding a solution to this problem is difficult and is returned to later in this chapter.

More generally, how the information is presented can affect respondent trust, where trust is important if the information is to be correctly construed. Analysis of the results of Burgess *et al.* (2000) would suggest that the manner the information was presented gave the impression of a hidden agenda and this affected trust in the questionnaire and the motives behind the study. If there is a lack of trust the respondent may refuse to consider the details. Overcoming a lack of trust may require sufficient detail such that respondents can trust the agencies responsible. Such a description may be impossible given the constraints of the individual questionnaire survey but may be more feasible within a group representation. In the absence of such detail respondents may well add their own meaning to the information presented, making it difficult for the researcher to interpret the responses given. Fischhoff *et al.* (1999) describe the opposite situation where respondents trusting the information provided may 'uncritically accept their initial impressions without clarifying terms' (p. 143). Clearly a balance is required between trust and blindly accepting the information provided. Table 5.3 demonstrates that the issue of trust was raised within most studies listed and is discussed in more detail below. The issue of trust is also returned to in Chapter 7, where during the post-survey groups respondents were observed to change their position on the issue and their resultant valuation responses.

Dealing with the information problem

The problems of presentation have been demonstrated. There are limitations in terms of time and cognitive ability of respondents within a questionnaire situation to take in, remember and construct their preferences. There is also the challenge of ensuring that the information presented is not misconstrued. The difficulty is the extent to which the valuations made reflect a consistent bundle of goods and services and pertinent information. Such consistency enhances the meaning of the response in terms of policy. However, as only a limited amount of information can be provided within individual surveys, respondents may not have sufficiently grasped the issues and, to compensate for this, may have filled the gaps in their understanding with their own interpretation of the scenarios considered. This suggests the importance of testing that the pertinent points have been grasped as the researcher can inevitably only provide a subset of information.

Through the use of exploratory group methods, Chilton and Hutchinson (1999) demonstrate how default assumptions can be explored and careful wording included (or excluded) where there are inconsistencies (consistencies) between the researchers understanding of the problem and that of the general public. Unfortunately Chilton and Hutchinson (1999) did not extend this analysis to explore how effective this expensive process had been in terms of improving the valuation made. Lazo *et al.* (1992), using a verbal protocol methodology, likewise adjusted information based on responses made and found the resultant values to have less variance and tests showing the valuations to be stable. Despite this promising result, the resultant information statement was 12 pages, suggesting a need to explicitly recognize that preferences are constructed and to provide very detailed information to the respondents.

Payment

The choice of payment vehicle has implications in terms of the implied property rights and who should pay for the scheme considered. It is strongly recommended that close consideration should be given to payment vehicle choice within the piloting of the questionnaire. Although it is unlikely that the choice of payment vehicle will be without controversy, a serious mismatch between the norms of the respondent and those implicit within payment vehicle choice may lead to responses being heavily weighted in terms of the rights and wrongs of the payment vehicle used, rather than the specifics of the environment scheme considered. Consequently, some respondents may protest and refuse to give an answer to the stated preference questions asked.

As a way of illustration, Vadnjal and O'Connor (1994) provide an example of a study where respondent attitudes were very deviant from those implied by the payment vehicle used. Vadnjal and O'Connor (1994) considered general public WTP to avoid development on Rangitoto Island in New Zealand. Rangitoto is an undeveloped volcanic island, which is densely vegetated and a notable landmark in the Auckland region. Although most respondents stated their WTP, the majority (77 per cent) considered their valuation not to be an accurate measure of the value of the island to them. This was clearly an extreme reaction and the use of follow-up interviews explored the cause of this problem. Basically, Rangitoto Island is very much part of the culture of the area and unspoilt enjoyment of the Island was seen to be a right of New Zealanders. This would suggest property rights are perceived to be with the general public rather than the developers, implying that the situation is one of compensation rather than a WTP. Vadnjal and O'Connor (1994) suggested the problems observed went beyond the issue of property rights where instead, the Island was considered not to belong to anyone.[10] As such, it is not an economic issue whether development should or should not occur on the island but rather a question of what is 'right and proper'.

Given the overriding misperception of the scenario, perhaps the above study should be seen as an investigation into public opinion regarding development on the Island rather than saying much about the validity of stated preference methods used. Clearly the valuation of Rangitoto Island provides such a conflict of social norms and values held by the general public that they are unable to make meaningful trade-offs between financial loss or gain and development on the Island. Despite demonstrating the inevitable limits of stated preference methods, the study did provide very interesting insights into the strength of feeling and the reasons behind such emotions. Perhaps this understanding would not have been gained without considering the potential trade-offs. However, with the benefit of hindsight, a group-based more social approach would have been more appropriate for exploring attitudes towards such trade-offs.

Having considered the extreme case of Rangitoto Island, more modest objections to the payment vehicle used have also been elicited from many of the other studies listed in Table 5.3. These studies provide an illustration of the range of problems that payment vehicle choice creates. The norm-activation model adapted by Blamey (1998) was found to be useful in explaining the comments made, particularly the denial of responsibility for payment and the denial of effective action (relating to trust in the authority responsible for delivery). These two issues are considered in turn.

Denial of responsibility for payment

The qualitative results show denial of responsibility to have been made in terms of the norm of fairness, where payment for the scheme was seen to be unfair due to the perception that the problems considered had been caused by others and/or other people are likely to benefit more. As such, the implication is that the responsibility for payment should be passed to others. A further related issue is that some respondents consider that they have already paid, through their taxes, rates, and so on, and it is an expectation that delivery on the scenario considered is something that should be expected as part of the perceived service contract. For an overview of comments raised, see column 4 in Table 5.3.

A denial of responsibility in terms of suggestion that 'others had caused the problem', was quite common in some studies. For example, Schkade and Payne (1994) suggest that the polluter pays principle was evoked for 13 per cent of the sample within their study. Likewise, Cameron (1997) noted that some respondents suggested polluters rather than the consumers should pay for the water quality problem considered. In the case of Svedsäter (2003) as many as 65 per cent of respondents suggested that those responsible should pay. Similarly, Blamey (1998) and Powe (2000) both reported that some participants blamed the government for poor decisions that have caused the problems being investigated. Again the implication is that the polluter should pay.

The schemes considered by Blamey (1998), Clark *et al.* (2000) and Powe (2000) benefit both farming and tourist industries. Within every focus group described by Powe (2000) concern was expressed that, rather than protecting the environment, the flood alleviation scheme would be defending land for farming or commercial holiday purposes. This issue appears to have been less important within the other studies. Blamey (1998) and Clark *et al.* (2000) merely noted the comments of one participant. Within the focus groups undertaken by Powe *et al.* (2006), concerns were raised by rural residents that urban residents would benefit sooner, and more, from street lighting improvement. This issue was not relevant to all studies, for example, Schkade and Payne (1994) did not note any complaints of an unequal distribution of benefits, but given that there are no clear commercial beneficiaries from protecting migratory birds, this was perhaps to be expected.

A further related issue is the spatial distribution of payment, where people living more locally to the schemes considered are likely to receive the most benefits. Indeed, Blamey (1998) noted denial of responsibility from individuals living outside the state of Southern Australia, within which the wetlands considered were located. Comments were raised that each state should deal with its own problems. Others saw the wetlands as a national problem and asset. Within the state of South Australia, concerns were raised that a precedent of national payment would be set such that other states would expect the same treatment for dealing with their problems. Powe (2000) noted that this later issue was discussed within two groups by respondents from in, and around, the wetland area considered. Clark *et al.* (2000) also notes similar problems with using a national payment vehicle to consider local issues.

Government taxation provides a further complication to this link, as, due to the way taxation is collected, for example, it is often unclear how revenue is spent. This situation can cause a lack of trust and lead respondents to suggest that they thought taxes were collected for 'these sort of schemes'. Such responses were common. Indeed, Blamey (1998) found that within every group meeting someone stated that responsibility for payment was with the government, with a typical response being 'what do we pay taxes for?' (p. 64). Only one participant mentioned this within the study by Powe (2000), which can perhaps be explained by the size and cost of the wetland flooding problem being considered beyond the normal service expectations. Schkade and Payne (1994) report that 12 per cent of respondents stated 'oil companies should pay', which may indicate a similar meaning to that observed by Blamey (1998). In one of the groups conducted by Powe *et al.* (2004a) there was a strong feeling expressed that biodiversity enhancement was a shared responsibility such that contributions from both the customers and the water company should be made. In fact, within the associated quantitative study the second most popular reason for non-payment was a feeling that the scheme is the responsibility of the water company. This issue was not noted within the other studies. The transcripts of Powe *et al.* (2006) also report this issue for two groups.

Perhaps due to the close correspondence between the water services provided and the payment for the good, Powe *et al.* (2005) report only one participant denying responsibility for payment. Everyone is used to paying for water supply and the environmental issues considered were only those directly related to water supply. Hence the link between who pays and policy outcomes was very strong. Also the scheme considered these issues for the whole of the water authority area, making a correspondence between payment vehicle and the spatial distribution of the environmental issues considered. The findings of Powe *et al.* (2005) suggest that the extent of the denial of responsibility problem depends on the nature of the problem being considered.

Lack of trust in the provider

Following the norm-activation model adapted by Blamey (1998) a lack of trust in the authority responsible for delivery can be interpreted in terms of a denial of effective action. Although the importance of the issue did vary between studies, trust in the authority responsible for provision was an issue relevant to most (Schkade and Payne, 1994; Vadnjal and O'Connor, 1994; Cameron, 1997; Blamey, 1998; Powe, 2000; Burgess *et al.* (2000); Svedsäter, 2003; Powe *et al.*, 2004a; 2005; 2006).

A lack of trust would appear to have sometimes been caused by concerns that the funding would not purely be used for the purpose stated. In a similar way to that described in the last sub-section, this may be partly caused by a lack of clarity as to how revenue, from taxation or water rates for example, is spent. Blamey (1998) and Powe (2000) suggest this situation has been made worse by experiences with road tax, where more revenue is collected than is spent on roads. Likewise, Cameron (1997) found allegations of previous misuse of funds to have caused feelings of lack of trust, where respondents did not trust the organization to spend the money wisely. Powe *et al.* (2006) found the key reasons for the lack of trust being; a concern that residents do not really see what they get from the local taxation they currently pay; whether the local authority will actually spend the money on the lighting scheme considered; and a strong perceived likelihood that the costs of the scheme will escalate beyond those stated. In the case of Powe (2000), there was a further complication due to a national taxation payment vehicle used for a local scheme. For example, one participant asked; 'is this [payment] solely for the Broads [name of the wetland area considered] or if it is being collected nationally would it be used elsewhere? Say there is a problem in Snowdonia?', and another asked, 'if there is a pressure from another source would they use it there?'. In the case of Powe (2000), there were also concerns that the revenue collected would not be used specifically on the flood alleviation scheme proposed but instead given to farmers or other commercial interests.

Further to issues relating to the actual authority responsible for provision, trust can be affected by questionnaire design. For example, Burgess *et al.* (2000) found mistrust, doubt and suspicion about how the funding would be used, which may have been partly caused by the particular focus of the presentation which left some respondents feeling that there could be a hidden agenda. This lack of trust led to other questions about the specific purpose of the survey; the rights given to those paying; enforcement of the payment; and the need for assurances that the finance raised would be used effectively.

It was noted above that a key issue is the lack of information on cost within stated preference surveys. The reason for its omission relates to the separate consideration of costs and benefits within cost-benefit analysis. Whether the absence of cost information affects trust is unclear, however a related issue is the use of the second bound when using the dichotomous choice CV method. The second bound is used to counteract the limitation of the information provided, where the analyst only knows whether a respondent is willing to pay above or below the price or bid level specified. As a consequence, practitioners have tended to favour a double-bounded approach,[11] where a positive response to the initial bid level determines a second, higher, bid level that is presented to the respondent (ascending sequence). Similarly, a negative response to the initial bid level determines a second lower bid level (descending sequence). This practice has been shown to significantly boost the statistical efficiency of the closed-ended or dichotomous choice approach (Hanemann *et al.*, 1991, Calia and Strazzera, 2000).[12] However, these efficiency gains are made at the expense of an increased propensity for bound effects and previous empirical research has suggested an internal inconsistency in the response strategies between the first and second bounds. This has led to the common observation of a lower WTP associated with the second bound responses (McFadden and Leonard, 1993; Carson *et al.*, 1994; Alberini *et al.*, 1997; Clarke, 2000; Bateman *et al.*, 2001; Powe *et al.*, 2006) and parameter inconsistency in the determinants of valuation responses (Cameron and Quiggin, 1994; Alberini *et al.*, 1997; DeShazo, 2002).

One possible explanation for this bound effect is the surprise of being asked the second valuation question, where respondents had the expectation that the first bid level was the price of the good considered. This can also induce a lack of trust in the agency responsible for provision, as illustrated using the qualitative results of Powe (2006) and Powe *et al.* (2006). For example, in response to being asked whether they felt pushed to say yes to the second higher bid level within Powe (2000) one participant stated: 'it's like a tramp asking for 5p for a cup of tea and then when you give him it he asks for 10p then he can have a roll as well'. Another participant stated:

> you are asking people to look at one figure then you might get a positive answer. If you say a fiver [£5] to look after the Broads then you say yes I'll do that. But if you then say if you are willing to pay a fiver will you pay a tenner [£10], suddenly human nature would say hang about your conning me. You wanted a fiver and now you want a tenner. Now come on how much is this going to cost?

A further participant made a similar comment: 'I start thinking you don't know what you are talking about. If you don't know what you are talking about you can't ask me for that money.'

Similar comments were also made by Powe *et al.* (2006), where the asking of the second bound was considered to be 'a bit sly', another 'a bit dodgy' and other participants suggested that the second bid made them question the realism of how much the scheme was actually going to cost and also led to participants feeling annoyed that it was asked. Within the main survey 59 per cent of respondents agreed with the statement 'if the council are unsure about the amount that we will have to pay for the lighting improvement scheme, it makes me worry how much this is actually going to cost'. It is clear that the use of the second bound within dichotomous choice increases the lack of trust in the questions being asked and as such reduces their validity. In the case of Powe *et al.* (2006) there was already a lack of trust in the agency responsible and the use of the second bound is likely to have made this worse.

Considering the implications, a lack of trust in the provider to use the money collected purely to implement the schemes considered would reduce the credibility of the transaction and may alter the valuations made. The key question here is whether respondents considered trust when answering the stated preference questions. Indeed, in two of the group meetings undertaken by Powe (2000), participants unanimously agreed that when answering the valuation question they assumed that the money would actually be spent on the scheme. A majority in another group agreed to the same principle. In fact, the spike model reported within Powe and Bateman (2004) suggest trust was not an issue within decisions as to whether people are willing to pay anything for the scheme or how much they were willing to pay.

Powe *et al.* (2005) present a similar finding of a lack of trust but having little effect on WTP. Here participants were asked 'if water charges were raised by your water company in order to finance the improvements stated, would you trust them to implement these schemes in practice'. The modal answer to this question was 'yes' (15 participants (46 per cent)), but ten participants (24 per cent) said 'no' and a further ten participants (30 per cent) said they did not know. Although the results are mixed, as with Powe (2000), the most important issue is whether participants assumed the money would be used for the schemes when answering the questions. In order to assess this, a further question was included and 25 participants (76 per cent) stated that they had made this assumption with a further seven (21 per cent) stating they had not. With the vast majority of respondents assuming the extra revenue to be used for the schemes considered, trust may not be a significant determinant of WTP within this study. Powe *et al.* (2004a) report a similar trust condition within the responses given.

For Powe (2000), Powe *et al.* (2004a) and Powe *et al.* (2005), WTP is unlikely to be affected by the issue of trust. However, within the actual implementation of the scheme the WTP stated is dependent on the authority responsible for provision actually demonstrating that the revenue will be used as promised. This need is illustrated through participant comments. For example, Powe (2000) reports the following participant comments:

> 'I'd pay as long as the people know where it is going.'
> 'if I had a 100% guarantee that that money was going to be spent on this scheme I would be in favour. I very strongly suspect that 99% of it would disappear off to something else.'

Such assurances were also requested within Burgess *et al.* (2000), Powe *et al.* (2004a), Powe *et al.* (2005) and Powe *et al.* (2006). Powe *et al.* (2004a) considers how people can be convinced within the actual delivery of the scheme.[13] However, what needs to be remembered here are the difficulties involved, or as one participant within Powe (2000) stated: 'convincing people that the money would actually go on the scheme would be very difficult'.

Within Cameron (1997) and Powe *et al.* (2006), trust was found to be a significant determinant of WTP. In the case of Powe *et al.* (2006) trust was a key determinant of both whether respondents were willing to pay anything towards the scheme, and also how much they are willing to pay. This lack of trust is illustrated in the attitudinal statement used within the main survey where 48 per cent agreed with the statement 'you can't trust local government to use council tax revenue to finance the street lighting improvement scheme'.

In conclusion, the results of the qualitative analyses considered demonstrate that funding adds an additional dimension to the meaning of the responses and complicates analysis. Although Arrow (1986) and Kahneman (1986) concluded that the public good nature of the goods and services valued using stated preference is likely to provide further differences than revealed when valuing private goods, gaining an understanding of these effects is difficult and raises questions of how the resultant valuations should be interpreted. For example, if WTP is significantly affected by trust, it could be asked if models should be calibrated so that the effect of trust is taken out of the valuations made. Clearly choice between payment vehicles will lead to different levels of trust, putting pressure on the research to choose correctly. Within Powe (2000) this issue was explored in detail and in six of the seven groups conducted a majority or all of the members agreed that national taxation is the most appropriate payment vehicle to provide the bulk of the money required. Despite this, attitudes towards the payment vehicle significantly affected whether people were willing to pay anything towards the scheme. The qualitative results also illustrated the confusion of some participants who misconstrued the payment vehicle as being local rather than national.

Motivations and Strategies Used

When faced with the actual stated preference questions, respondents generally found them difficult to answer. For example, Powe (2000) suggests that the CV questions were generally thought to be the most difficult on the questionnaire, Svedsäter (2003) found ten respondents (35 per cent) to claim the stated preference questions were impossible to answer; and it was evident from the work of Clark *et al.* (2000) that participants were 'struggling with the money business'. Also in terms of choice experiments, Powe *et al.* (2005) suggested most participants found the choice task difficult. Having considered the presentation and payment vehicle, this sub-section will explore respondent motivations and strategies used to overcome the difficulties encountered by respondents when answering stated preference questions. Some of the motivations and strategies used are consistent with economic theory and expectations. However, others illustrate the use of simplifying heuristics which, may or may not be consistent with economic theory, but do add complexity to developing an understanding of the meaning of the responses made. Table 5.4 provides a summary of the motivations and strategies used when answering stated preference questions.

Table 5.4 Comparison of qualitative studies (strategies for answering stated preference questions)

Study	How much could afford	Opportunity costs of payment	Charity like comments	Symbolic for a broader good	Guessed / made up	Fair share / bid realism
Schkade and Payne (1994); Schkade and Payne (1993)	Income (p=0.06), mentioned 31% (p=0.77); tax increase not noticed (5%); use and price of petrol (23%) (not sig)	What has to give up to pay 0%; 74% willing to support; and average 2.9 other causes	Charity/good caused 17% (p=0.00)	Broader env. concerns 23% (p=0.00)	Guessed/ made up 20% (p=0.49)	No. households/ fair share 41% (p=0.00)
Fischhoff *et al.* (1993); Fischhoff *et al.* (1999)	Not tested	Not tested	Not tested	Not tested	Not tested	Not tested
Vadnjal and O'Connor (1994)	Not tested	Few stated opportunity cost	None stated	None stated	None stated	Some asking who else is going to pay
Cameron (1997)	70% 'change in personal circumstances' explained ΔWTP	Non stated	More a donation to a worthy cause	Symbolic meaning for the problem of society	Not stated	Not stated
Burgess *et al.* (1998); Clark *et al.* (2000); Burgess *et al.* (2000)	11 stated how much could afford; two suggested that the bid amounts were small	One considered substitutes; four struggled with the context of the question	One comparison with other charities	Three broader good	Two just came up with a value	One aggregated amount based on the number of people likely to pay
Blamey (1998); Blamey *et al.* (1999)	Low price – not miss financial outlay	More likely to be stated by those living further away from the wetland area considered	None stated	Some participants broader good and doing their bit	None stated	Some suggested would pay what is reasonable and others guessing the cost and number paying

Brouwer *et al.* (1999); Powe (2000); Powe and Bateman (2003, 2004); Bateman *et al.* (2001)	Income (p=0.00), five out of seven groups considered comments that the amounts were small	Discussed in two groups	Discussed in two groups	Discussed in one group	Not discussed	Discussed in every group, majority considered during interview
Fischhoff *et al.* (1999); Welch and Fischhoff (2001)	Not tested	33% considered payment for status quo, 40% other pollutants in river	Not tested	Not tested	Not tested	Not tested
Svedsäter (2003)	41% could afford	28% otherwise spend on taxes	10% charity like	35–55% environment in general	10% guessed / made up, 10% uncertain of answers	10% what is reasonable, 28% will pay what it takes
Powe *et al.* (2004a)	Discussed in all groups, three groups small amounts	Discussed in context of other services – three groups	None stated	Discussing in two groups	Discussed in one group	None stated
Powe *et al.* (2004b), Powe *et al.* (2005)	Discussed in five groups, some saying would not notice payment	None stated	None stated	None stated	None stated	None stated
Willis *et al.* (2005); Powe *et al.* (2006)	Discussed in two groups	Other services discussed three groups	None stated	None stated	None stated	None stated

As suggested by Fischhoff *et al.* (1993), the use of concurrent verbal protocols elicited within individual interviews is the most appropriate method of assessing the motivations and strategies used. Using a retrospective approach, respondents may not correctly recall the factors considered during valuation. Group-based approaches create different conditions from that of the actual valuation situation, where the latter may lead to a greater discussion of social issues rather than the more private concerns of individuals. Despite these valid concerns, consistency across different methods still provides useful information.

How much could be afforded

Consistent with the findings of most quantitative studies, the results of the qualitative analysis suggests that a number of respondents considered their personal circumstances and how much they could afford (Schkade and Payne, 1994; Cameron, 1997; Blamey, 1998; Clark *et al.*, 2000; Powe, 2000; Svedsäter, 2003; Powe *et al.* 2004a; 2005; 2006). For example, Schkade and Payne (1994) noted that 31 per cent of respondents mentioning 'family income/expenses' and Svedsäter (2003) found 41 per cent of respondents to have considered what they are able to afford. In addition, in five out of the seven group meetings, Powe (2000) notes participant comments suggesting they had considered how much they could afford; and Clark *et al.* (2000) noted that all participants considered their individual financial circumstances and whether they could afford the amount. Furthermore, Cameron (1997) notes that a change in personal circumstances was the key issue for changes in WTP within their longitude study. The second column in Table 5.4 provides a summary of the key issues raised within the qualitative studies relating to how much people could afford.

It would appear that some participants did consider whether they could afford the amounts stated, however, perhaps due to the hypothetical nature of the survey some respondents may be agreeing to pay with provisos. For example, Powe (2000) found three participants in separate meetings to have agreed to pay the stated amount but then suggested due to their current financial situation they could not actually afford the payment. Similarly, Svedsäter (2003) reports one participant stating that: 'when I would actually have to reach for the money and pay, I wouldn't be that happy. So take that with a little bit of reservation' (pp. 130–1). Clearly some respondents answer differently due to the hypothetical situation, but there is also as much evidence from other respondents that they had seriously considered their current financial situation (Powe, 2000; Powe *et al.*, 2006). Overall there is likely to be a positive bias, the magnitude of which is difficult to determine from the qualitative responses. Recent work by Cummings and Taylor (1999) and others have demonstrated how this bias can be reduced by encouraging respondents to consider their personal financial situation.

There is evidence to suggest that where bid amounts or prices are stated, they are sometimes considered to be too small to sufficiently test how much respondents are willing to pay (Schkade and Payne, 1994; Blamey, 1998; Clark *et al.*, 2000, Powe, 2000; Powe *et al.*, 2004a; 2005). For example, the transcripts from Powe *et al.* (2004a) report three group meetings where the size of the amounts were discussed, where one participant described the amounts to be 'trivial' and another stating 'to be honest I wouldn't notice'. In one group, however, one participant stated if 'every company did this then you might notice then', and in another one participant noted their inability to pay due to their financial situation. In the case of Powe (2000) in five out of seven meetings participants suggested that they were not tested by the amounts, for example, one participant stated 'the specific amounts were so low £5 and £10, I wasn't tested to how much I was willing to pay' and in another group 'it is so easy to answer yes if you get a lower figure'. As noted above, the problem here is that by combining the task complexity and the hypothetical nature of the scenarios, there is a need not only for respondents to take the amounts serious, but also that the price is sufficient for respondents to put in the required effort to consider the full implications of the payment. Powe (2000) illustrates a respondent not actually putting in sufficient effort because the amounts were small: 'Had I been asked for £100 or £150 I probably would have thought about that [the valuation scenario considered] a lot more than I did but the maximum one on here was only £20 and I was quite prepared to pay that'. Furthermore, it was suggested by one participant that at the lower bid levels the actual amount may not be meaningful.

Opportunity costs of payment

Column three in Table 5.4 lists the extent to which respondents consider opportunity costs of payment when answering stated preference questions. Perhaps the simplest form of opportunity costs to consider are those relating to the personal sacrifice of the individual in terms of other goods and services forgone. There was a general lack of consideration of the opportunity costs of payment. For example, Schkade and Payne (1994) report no respondents considering what they would have to give up to make the payment, and Blamey (1998) notes only one respondent considering what they could otherwise buy with the money. Furthermore, Powe (2000) reports that this issue was discussed in two groups and Svedsäter (2003) reports 14 per cent of respondents considering the impact on other expenditure.

Further to their personal circumstances, it is also hoped respondents will consider opportunity costs in terms of other schemes that the respondents could spend this money on. A failure to consider preferences for the scheme considered in the context of complements and substitutes may lead to systematic bias, where undue priority is given to the specific scheme considered. Given the concerns expressed within the payment vehicle sub-section previously regarding a use of funds on other schemes, respondents would at least appear to consider alternative uses of the funds. In the case of studies where a local scheme was considered with a national payment vehicle there seemed to be an awareness of the alternative schemes, but awareness for some respondents would appear to have developed within the group meetings rather than something considered whilst completing the valuation questions. Powe (2000) reports that substitutes were mentioned in two of the seven groups undertaken, however, the evidence suggested that this was sometimes an afterthought (see Chapter 6 for details), perhaps suggesting that using the focus group approach the respondents have more time and are more able to consider the wider issues that need to be considered.

Schkade and Payne (1994) did not list awareness of substitutes as an issue considered by respondents. Instead, substitutes would appear to have only become apparent to the respondents during a follow-up question; 'how many other important environmental issues would you agree to support with a similar dollar amount each year?' Indeed, despite the thousands of worthy causes, 74 per cent of the respondents said they would support an average 2.9 other causes. When confronted with the range of other causes Schkade and Payne found several respondents then realizing the far reaching implications for their household budget and indicated that 'the amount they stated was really too large or that it should go for all similar issues' (p. 102).

It was clear that substitutes had been considered by participants/respondents within other surveys. For instance, Clark *et al.* (2000) report a discussion of substitutes, where, for example, one participant assessed how the area compared to other nature conservation sites. Another participant, however, stated they were not qualified to judge the worth of a particular scheme in relation to the many others. Blamey (1998) noted that some participants considered substitutes and that this was more apparent the further participants lived from the study site. The transcripts from Powe *et al.* (2004a) suggest substitutes were considered in all groups, where these related to other goods and services provided by the company. Likewise, the transcripts of Powe *et al.* (2006) report that the scheme was considered in the context of other services provided by the local authority within three groups, where in one group there were very strong opinions expressed that there were higher priorities than street lighting.

As substitutes are integrated within choice experiments, instead the important validity issue is whether a relevant substitute has been omitted. Given the wide remit of the questionnaire used by Powe *et al.* (2005) this was not found to be a problem. As trade-offs between alternative schemes are implicit, it was interesting to explore the ease or difficulty of the task. Powe *et al.* (2005) report that participants found the trade-off between environmental quality, service and cost relevant and most responses reflected a balance between these issues. Some participants stated that they could choose between environmental attributes, and a statistically significant difference in preference was observed. However, participants generally found such choices more difficult than merely trading-off between the environment, service and water charges. Indeed, some thought they had inadequate knowledge and experience in order to make valid responses. This demonstrates the difficulties encountered in considering substitutes even when trade-offs are explicit within the survey. The implications of these findings for CV, where alternative schemes are not explicitly stated, are that the ability to give proper consideration to substitutes is likely to be limited.

Charity like comments – simplifying heuristic

Given the novelty of the elicitation methods used, respondents may search for something similar or a reference point to help them decide on their response. As environmental issues can be regarded as 'good causes', some respondents may consider the situation to be like that of donating to a charity. Charity like responses may give respondents a warm glow or moral satisfaction when responding (Kahneman and Knetsch, 1992), where respondents may or may not consider there to be a link between response-policy-payment and as such their budgetary constraint.[14] Baron and Greene (1996) have considered moral satisfaction to consist of two components; contribution; and warm glow. Using the contribution explanation, respondents view the transaction in a similar way to donating to a charity. Responses may still be consistent with ability to pay, but also reflect their willingness to 'do-their-bit' for environmental concerns in general, rather than the specific good considered. Such personal participation may also give the respondent a warm glow value (Andreoni, 1990). In the case of the contribution form of moral satisfaction, those respondents will be aware of their budgetary constraint and are likely to state a lower WTP because such respondents may consider substitute charities to which they could donate. Where respondents do not feel they will actually have to pay the amount, this may lead to a positive bias in responses.

Considering the evidence, Schkade and Payne (1994) found 17 per cent of their sample to have made charity like comments, where, for example, the amount stated was equivalent to amounts given by the respondent to charities. Confirming the contribution form of moral satisfaction, such respondents stated a significantly lower WTP than others at the 5 per cent level. More generally, Clark *et al.* (2000) noted charity like responses, where two participants, for example, refused to pay because they were uncertain what other charitable demands there might be. Powe (2000) also reported 'charity-like' comments to be made in two of the group meetings. In one group, a comment was made that the bid levels used should be equivalent with what people give to charities. In another group one participant stated:

> it's like picking out a favourite charity I mean there are so many charities who do you give to, you can't give to them all. At least you could try to give to them all but if you did you would end up giving a penny to each one. So what you do is pick out two or three charities and give a pound to you, a pound to you and a pound to you.

This quote suggests a similar finding to that of Schkade and Payne (1994) made above, where respondents were observed picking only a few causes to support in order to 'do their bit'. Cameron (1997) also noted some respondents suggesting charity like responses.

The study by Powe *et al.* (2004a) provides results that are both unusual and interesting. Within the main survey by Powe *et al.* (2004a), the respondents were not asked to consider their WTP but instead their willingness to forgo a potential bill reduction in order to fund biodiversity conservation. Within the focus groups the participants were then asked to consider the situation of a bid increase. Generally, there was a negative response to the idea of a bill increase, with participants in two groups comparing the situation to a charity and suggesting that they would not view their water authority as their favourite charity. As an illustration one participant stated:

> I would be doubtful that the water company was the right place for my money to go, because they are a water provider. I would look to some of the organizations that have been mentioned so far [Royal Society for the Protection of Birds and other environmental charities].

These results may suggest that forgoing a bill increase would not be considered as contributing to a charity.

'Charity-like' responses were not noted by Blamey (1998), Powe *et al.* (2005) or Powe *et al.* (2006). In the case of the latter the scenario was unlikely to be viewed as charity like for a street lighting scheme for the whole of the local authority area. As such the extent of the charity problem will depend on the type of goods and services being valued.

Symbolic for broader good – principled response

If the motivations and strategies are to be consistent with expectations, there is a need for respondents to consider the specifics of the scenarios considered. Although there is evidence to suggest that respondents consider scenario specifics, there is also evidence to suggest that others pay little attention to the extent of scope of the scheme considered. For example, in the case of environmental goods respondents are often motivated by a need to do as much as they can. As such, they may consider their ability to pay and if the particular environmental problem considered is worthwhile and then state the maximum they are willing to contribute towards solving the particular problem. This leaves the issues of how the environment will be protected or improved and the extent of the change, for example, in terms of the number of birds saved or area of land protected, to the agency responsible for delivery. For some respondents they may fail to consider any specifics of the goods and services valued, but instead see it as having a wide symbolic meaning for a broader good. As this issue relates directly to scope sensitivity and will be discussed in detail within Chapter 6, only a few summary details are provided here.

Schkade and Payne (1994), Blamey (1998), Cameron (1997), Powe (2000), Svedsäter (2003) and Powe *et al.* (2004a) all provide evidence that respondents see their WTP as symbolic for a contribution towards solving environmental problems in general. The magnitude of the problem, however, varies between studies. For example, Schkade and Payne (1994) report that 23 per cent of respondents related their response to a broader environmental concern than the specific scheme considered. These respondents were found to state significantly higher WTP amounts at the 5 per cent level. Blamey (1998) also noted that some individuals were not concerned with the specifics of the scenario considered, but instead with 'attitudes and beliefs of a more general environmental nature' (p. 60). Clark *et al.* (2000) found that a number of participants considering the scheme as symbolic for broader problems of, for example, the reduction in diversity of nature. Perhaps due to the focus groups only considering people living in and around the area of the wetland area considered, with a few exceptions, Powe (2000) notes there was little evidence of respondents considering their response as symbolic for environmental goods in general. Cameron (1997) also noted some respondents suggesting their responses were symbolic for wider societal problems.

These findings beg the question of whether symbolic/broader good responses are a symptom of the 'good cause' of environmental protection and improvement. Such broader meaning is easy to attach to biodiversity as it is consistent with the idea of 'think globally, act locally'. Indeed, within two of the groups reported by Powe *et al.* (2004a), there was a discussion relating to respondents doing their bit locally to help solve the global problem. However, where environmental issues were considered purely in the context of water supply delivery, Powe *et al.* (2005) report no respondents suggesting symbolic reasons for their choice between environment, water supply and water rates. In the non-environmental study by Powe *et al.* (2006) there were also no comments relating to a broader good. This is perhaps due to the nature of the good considered; street lighting. However, due to the lack of trust of the agency responsible for provision and the effect this had on WTP, a number of such responses are likely to be symbolic for the perceived general poor performance of the local authorities involved. As such symbolic meaning can also be attached to the payment vehicle.

Guessed/made up

An indicator of the accuracy of the valuation responses is the extent to which respondents consider themselves to have guessed or to have just made up their responses. This may indicate both the cognitive effort required when answering such questions, but as you are less likely to guess with a real payment, such a response may also reflect the hypothetical nature of the payment. A reassuring finding was that only in the case of Schkade and Payne (1994), Clark *et al.* (2000), Svedsäter (2003) and Powe *et al.* (2004a) did respondents/participants note that they had guessed or made up their responses. The degree of guesswork varied between studies, but in all cases there were other more important determinants of WTP. This might be partly related to the elicitation method used, as the open-ended CV approach is more prone to guesswork and may also be more difficult for respondents to answer. Using the dichotomous choice approach, for example, there is little scope for the respondent to guess/make up a value.

Considering the evidence, Schkade and Payne (1994) found 20 per cent of the sample to have just guessed or made up a number, but as such a response was not significantly related to WTP, the effect of this uncertainty did not have a biasing effect. Such responses are unlikely to accurately reflect the respondent's value and suggest they lack coherent preferences for the goods considered. Clark *et al.* (2000) noted that it was evident that their participants had put a lot of effort into answering the CV question however two participants stated 'they had just come up with a number'. Svedsäter (2003) also noted a small number of respondents saying they had guessed when responding. From the transcripts of Powe *et al.* (2004a), in one group some participants gave the impression they had guessed when answering.

Fair share / bid level realism

A key motivation within stated preference studies is to develop an incentive compatible elicitation method such that respondents state their maximum WTP. However, stating their WTP for public goods that have communal payment vehicles, a primary concern of respondents may be that they should only pay their fair-share.[15] If such a 'fair-share' heuristic were used, the elicitation would not be incentive compatible and would reflect, instead, the perceived cost per household of the scheme rather than the value of the good to the respondent. The open-ended version of CV is particularly susceptible to the fair share heuristic as no guidance is given as to the price of the good (Schkade and Payne 1994; Bohara *et al.*, 1998). If it is assumed by the respondent that the bid level reflects the actual price upon scheme implementation and their fair share of the cost, this problem may not be relevant using the dichotomous choice method (Hanemann, 1994). Despite this, Stevens *et al.* (1994) has reported the use of the 'fair-share' heuristic within dichotomous choice responses. In the case of dichotomous choice surveys where a specific local issue with national relevance is valued using a national payment vehicle such as used by Blamey (1998), Clark *et al.* (2000) and Powe (2000), cost information would make all but the smallest bid levels appear unrealistic. This could lead to respondents not taking the questions seriously enough and/or thinking they would not actually have to pay as much as the bid level.

Considering the evidence, the issue of paying their fare share would appear to have been very important within the studies of Schkade and Payne (1994), Blamey (1998), Powe (2000) and Svedsäter (2003) and has been noted within the analysis of Vadnjal and O'Connor (1994) and Clark *et al.* (2000). In terms of the specific studies, Schkade and Payne (1994) found 41 per cent of their sample to suggest that: 'if everyone did their part then each household would not have to give all that much' (p. 99). This was the most frequently stated issue reported by Schkade and Payne (1994) with respondents mentioning this issue stating a significantly lower WTP at the 5 per cent level. Likewise, Svedsäter (2003) reports 10 per cent of respondents suggesting they should pay what is reasonable and 28 per cent stating they would pay what it takes. As Svedsäter (2003) considered global warming the communal nature of the problem was clearly evident.

The use of the fair share heuristic was similar within both Blamey (1998) and Powe (2000) and in the case of the latter was an issue raised within every group meeting. Within one group reported by Powe (2000), further to the majority considering what they were able to afford, the group unanimously stated they considered the costs of the schemes whilst answering the questions. Whether consideration of cost led respondents to question the realism of the bid levels asked seemed to depend on their size. For example, a lady offered the £10 bid level stated 'I assumed these figures had been arrived from people doing their sums' and a further participant suggested 'you've got these figures £10–£15 as a general guess as to what it might cost to do this'. Hence, these findings may question the validity of the higher bid levels but not the lower.

Powe (2000) also reports several statements suggesting that respondents really did not know how much the flood alleviation scheme would cost. One participant stated: 'I couldn't put a cost on it, how many billions of pounds would it cost?' and another referring to the bid levels and the flood alleviation scheme considered, 'I don't think that will be enough however much we spend we will still need to spend more'. Furthermore, in another group one participant stated: 'if someone did cost a project like this then it would triple and quadruple after they started doing it'; and a further participant stated: 'if everyone in Broadland [the wetland area considered] had to pay £200 it would only be a minute amount in comparison to the costs of a scheme like this'. Referring to the latter comment the truth is somewhat different, with there being somewhere in the region of 30 million households in Britain and the scheme was at the time estimated to cost only in the region of £70 million.

Powe (2000) also reports evidence of participants questioning the realism of the bid levels. For example, one participant stated: 'mine [bid level] was so high [£500] I just thought "gulp", if you are looking round the entire country, £500, then that's a lot of money', and in another focus group a respondent stated: 'this [the bid levels] didn't make sense if you multiply your £100 by 20 million for the number of households'; and a further stated 'how much is the whole thing expected to cost? What I was thinking was if it is going to be paid through your taxes you just have to divide the number of tax payers and you can come up with a figure.' Despite these findings, it was not clear whether they were retrospective thoughts or those considered during the completion of the questionnaire.

In conclusion, the evidence suggests the use of the fair share heuristic is a social norm that is widely held when valuing 'good cause' public goods using communal payment vehicles and both the open-ended and dichotomous choice CV studies are susceptible to this problem. In the case of the open-ended method use of the fair share heuristic results in a downward bias which has been empirically tested (Schkade and Payne 1994; Bohara *et al.*, 1998). In the case of the dichotomous choice approach the heuristic would still appear to be evident but only in the case of bid levels being considered to be realistic should it have an effect on WTP. Unfortunately, this effect has not been tested empirically. If the problem is found to be significant then it would suggest a more social approach to estimating WTP where groups rather than individuals would be asked to determine a societal WTP that everyone should pay. This issue is considered in more detail within Chapter 8. Interestingly, using the choice experiment approach and price attribute levels within the 'ball park' of the actual costs, the transcripts of Powe *et al.* (2005) state that the participants unanimously agreed that the amounts were assumed to reflect the costs. Hence in the case of Powe *et al.* (2005) at least the fair-share heuristic was not a problem.

Summary and Discussion

Due to the length of this section it is necessary to summarize. This section has considered the issues of presentation, payment vehicle choice and the motivations and strategies used by the respondents when answering valuation questions. A number of positive findings have been stated, however, at least as many complications have also been identified.

The consideration of the information problem has highlighted the constraints of time and cognitive ability of respondents within a questionnaire situation. Although a level of information can be provided that most are satisfied with, respondents may also misconstrue this information, either through their misunderstanding or mistrust of the information presented. The implications of payment vehicle choice were also far from simple. Indeed, it was demonstrated that an 'ideal' payment vehicle might not always exist, where even the choice of the payment vehicle that is most popular with the respondents might still lead to problems of respondents denying responsibility for payment; lack of trust in the authority responsible for provision; and confusion as to the context of the payment. Lastly, a sizeable number of respondents would seem to consider their ability to pay and a smaller proportion would also appear to consider the opportunity costs of payment. However, problems of respondents considering the stated preference scenarios to be 'charity like' (negative bias) and symbolic for a broader good (positive bias), may reduce the meaning of the valuations elicited. Furthermore, given the public good nature of the goods valued and the communal nature of the payment, many respondents follow the social norm of paying their fair share, rather than giving responses consistent with their maximum WTP.

GENERAL PUBLIC REACTION TO SURVEYS

Further to exploring the complexity of survey design and interpreting the meaning of valuation responses it is also important that insights are gained into the public acceptability of the approach and the acceptability of the valuations within policy making. Given the controversial nature of environmental valuation this additional information is very important in determining the policy relevance of the information provided. Cameron (1997), Powe (2000), Clark *et al.* (2000), Welch and Fischhoff (2001), Svedsäter (2003), Powe *et al.* (2004a) and Powe *et al.* (2005) have all considered the acceptability of the valuation approach to the general public. The results are mixed with Clark *et al.* (2000) finding the approach to be largely unacceptable, whereas the results of the other studies are less conclusive.

Beginning with the study consistently raising the most concern, Clark *et al.* (2000) report that the participants 'unequivocally rejected CV as an acceptable way of representing their values, or views, to decision makers' (p. 60). In fact, local residents suggested an inability to put a monetary value on something you feel passionate about. They also expressed feelings of moral outrage at the use of a monetary sum to measure what individuals saw as their ethical and moral values. More generally, there was a consensus in all three groups that decisions about local environmental schemes should be made by the government, advised by experts and in the context of other nature conservation schemes. Concerns were also stated about the need for local knowledge and values to be communicated to decision makers but participants generally felt that stated preference was not an appropriate mechanism for achieving this.

Svedsäter (2003) found 45 per cent of respondents to claim that the environment is not a monetary issue and that policy making should not be based on 'private economic decision making' (p. 132). As this percentage suggests, some respondents were in favour of CV arguing it is a 'sensible and feasible approach, that policy makers should take account of public opinion, and that economic benefits and costs matter' (p. 132). The balance in favour of the use of the CV approach was stronger within the study by Powe (2000) where in five of the seven groups a majority considered the overall approach acceptable and suggested that the answers were meaningful and accurate enough to inform actual decision-making.

In terms of choice experiments, Powe *et al.* (2005) report that 61 per cent of respondents suggested that their responses to the stated preference questions are sufficiently accurate to guide policy decisions on water supply and only 8 per cent not. However, following the discussion, three participants switched from 'yes' to 'no' so that only 55 per cent of the participants suggested their stated preference responses sufficiently accurate to guide policy decisions and 14 per cent stated that they were not. Participants were prompted to explain why they responded as they did. Comments from those suggesting their responses to be sufficiently accurate to guide policy explained that was why they were answering the questions; that they were being asked to guide policy not to make decisions; and that their support was provisional on there being an interview situation in which they could concentrate. Those not liking the approach suggested that environmental decisions should be made by experts; based on the information given they could not choose; and that they did not trust that their responses would be used appropriately. Explanations for the 'don't know' responses were a lack of understanding as to why they were being asked; and a disbelief that they would be listened to.

When asked how participants would react if they were to find out that their water company had increased their water charges in order to fund environmental improvements; Powe *et al.* (2005) report the reaction of most participants to be cautious, showing concern that the bill would increase by only the order of magnitude stated in the questionnaire, and that they would like to know/see what has been done with the money. Although some negative comments were made about water companies, subject to the two caveats stated, participants were generally happy with the proposition. Using the same approach, Powe *et al.* (2004a) note a similar reaction in two of the groups, but in one of the groups there were strong objections voiced by some participants regarding the use of the results of stated preference surveys in this way.

Welch and Fischhoff (2001) examined the impact of explaining to respondents about the CV method either prior to conducting the valuation exercise or following the valuation exercise. Although the majority were still favourable to its use, the results showed that the more respondents knew the less happy they were about the application of CV within policy making and the less willing they were to participate in future CV surveys. Although information about CV and its use within policy making did affect 18 per cent of the WTP responses, changes were balanced in both directions and there was no overall effect on the valuations made. Interestingly, those rating the CV approach more highly also stated a higher WTP.

Cameron (1997) reported mixed findings, suggesting there to be a polarization of opinion. Those supporting the use of CV held the view that water quality problems were eminently solvable by technical means and economic signals. Whilst opponents justified their position by suggesting that more money would not solve the problems and there was a need for a fundamental change in the relationship between humans and the environment. Here, CV can be seen to be appropriate when considered from an 'instrumental' viewpoint, but not from other perspectives. However, again objections would appear to be specific to the problem considered and to the solution being suggested, where questions were raised as to whether the scheme was appropriate and whether the 'right' result would come from the use of CV. After three meetings over as many years, Cameron (1997) found the majority to be in favour of the use of CV for the relevant policy considered.

Although the balance of opinion regarding stated preference methods varied between studies, the issue of complexity and the need for expert knowledge was a recurring theme. For example, Svedsäter (2003) suggested a common argument raised was that lay people do not have sufficient knowledge of the issues and that experts should decide. Given that the scenario considered was global warming, with its complexity and uncertainties, it is hardly surprising that some respondents held this viewpoint. There was also the perceived inappropriateness of considering the issue within a given nation, where a view was stated that the problem should be decided through joint efforts across nations. Furthermore, Powe (2000) reports that in the two groups where a majority considered the overall approach not to be acceptable, the main reason given for this position was that the scheme considered and its consequences were too complex to be able to make a decision on it in the available interview time or without expert knowledge. Within the studies by Blamey (1998), Powe *et al.* (2005) and Powe *et al.* (2004a) issues of insufficient understanding of respondents and the need for expert knowledge were also raised. This leads to issues of representative democracy rather than a more participatory form applied using stated preference methods.

Overall the results would suggest that it is perhaps inevitable that there will be dissenters, but for some scenarios it is possible to gain a majority in favour of stated preference, even when informed in detail of the method being used and its proposed use within policy making. These results have suggested however, that it is not sufficient for decision makers to simply ask for the value the general public place on a scenario considered, but they also need to ask what kind of values they are able/willing to articulate and whether the method is appropriate for the scenario being considered.

CONCLUSIONS

Through the insights gained from the use of qualitative methods, this chapter has considered how to interpret stated preference responses. In order to provide a framework for this analysis, the fundamental problems have been considered in terms of those identified through the lens of economics, behavioural psychology and more social/institutional approaches. Economists tend to focus on the need to design transactions that are incentive compatible, where there is an expectation that such a characteristic will encourage meaningful responses. However, psychologists focus on the constructed nature of stated preferences, where the inevitable complexity of the scenarios considered leads to questions about the meaning of the valuations elicited. From a more social/institutional perspective, social norms and commitments are seen to guide responses, further complicating their interpretation.

The qualitative findings of this chapter have been presented in terms of issues relating to survey design (presentation, payment and transaction mechanism), strategies used by respondents and the public acceptability of the approach. Throughout the analysis, the complexity of issues has been evident, where interpretation requires the application of additional resources for qualitative analysis and the use of detailed attitudinal statements. Due to the presence of such complexities and the potential for misinterpretation, a failure to employ such resources may lead to the incorrect application of values within policy making.

The findings of the qualitative analysis and their implications for stated preference methods are now considered, where the issues of cognitive limitations, hypothetical payment and the communal nature of the scenarios and payment vehicles are all seen to have relevance.

Presentation

Difficulties encountered within the presentation of the scenarios (including payment) have demonstrated the inevitable limitations in terms of the time to digest the information and the complexities of the scenarios considered. Although a level of information can be provided that most are satisfied with, respondents may also misconstrue this information, either through their misunderstanding or mistrust of the information presented. Such problems question whether respondents can adequately construe the pertinent information and whether there can be sufficient consistency within respondent understanding such that the valuations are meaningful. Whether the incentives within the stated preference questions were adequate for respondents to put in sufficient effort is unclear. There is evidence however that many respondents did not feel tested by the small size of the monetary amounts and this may have led to the superficial understanding observed within some of the studies. Where group meetings have been used, this understanding was increased. The extent to which the use of group methods provides a solution to this problem will be investigated in Chapter 7.

Payment Vehicles

The results of the qualitative analyses demonstrate that funding adds an additional dimension to the meaning of the responses and complicates analysis. Indeed, it was demonstrated that an 'ideal' payment vehicle might not always exist, where even the choice of the payment vehicle that is most popular with the respondents might still lead to problems of respondents denying responsibility for payment, a lack of trust in the authority responsible for provision and confusion as to the context of the payment. Many such problems were due to the communal nature of the payment and the lack of transparency within which the funds from such payment are used. Although such findings may be consistent with economic theory, it is difficult to gain an understanding of these effects and interpretation of the valuations is complex.

Strategies and Motivations

Whether following a social norm of honesty or due to a perceived link between response and policy, a sizeable number of respondents would seem to consider their ability to pay and a smaller proportion would also appear to consider the opportunity costs of payment. However, problems of respondents considering the stated preference scenarios to be 'charity like' (negative bias) and symbolic for a broader good (positive bias), may reduce the meaning of the valuations elicited. Furthermore, given the public good nature of the goods valued and the communal nature of the payment, respondents have a tendency to follow the social norm of paying their fair share, rather than giving responses consistent with their maximum WTP. In summary, these responses strongly suggest that ability to pay, price, quantity and quality only provide a partial understanding of the strategies and motivations used when responding to stated preference questions; with social, cultural and ethical factors also playing their part. As such the resulting behaviour is much more difficult to predict and generalize about than neoclassical economics assumes.

Acceptability of the Approach

The qualitative results have demonstrated it is inevitable that there will be respondents that object to the stated preference approach, particularly in terms of environmental goods. However, the evidence also suggests that for some scenarios it is possible to gain a majority in favour of the use of stated preference, even when informed in detail of the method being used and its proposed use within policy making. These results have suggested however, that it is not sufficient for decision makers to simply ask for the value the general public place on a scenario considered, rather, instead, they also need to ask what kind of values they are able/willing to articulate and whether the method is appropriate for the scenario being considered.

Overall these results illustrate the complexity of valuing environmental goods. The qualitative results are mixed and interpretation of the implications for stated preference methods will depend on the researchers' attitude towards the ability of respondents to articulate their preferences and the need to make decisions in a social setting and/or with real economic payments. Whatever the researchers leaning regarding these issues, the results also suggest that the problems identified can be reduced through good design, selective choice of scenarios that are not too complex, and with which respondents have a high level of engagement, and/or have consistency between the payment vehicle and the scale of the goods and services valued. The degree to which these issues can be met will affect the quality of the valuations elicited. Consistent with the argument generated throughout this book, if stated preference surveys are going to be undertaken, the use of qualitative methods is essential and needs to be integrated within the method adopted in order to improve design and help interpretation of responses.

NOTES

1. Qualitative methods can also be used to test the sensitivity of responses to deliberation. This is absent from this list because it is discussed in Chapter 7.
2. Framing is also known as a violation of the logic of extensionality (Arrow, 1982).
3. See the next chapter for a detailed discussion of this issue.
4. The debate over the sensitivity of estimates to the scope of the good considered provides an example where split-sample comparisons have commonly been made between the WTP (Carson *et al.*, 1998). This issue is considered in detail in the next chapter.
5. For example, Sagoff (1988) introduced the dichotomy between citizen and consumer when discussing the different roles that people perform within their life. For instance, in the professional role a person may wish to visit the office at the weekend, whereas, in a parental role they may wish to spend time with their children. The form of stated preference questioning can focus on consumer aspects of the problem.
6. A similar situation could exist when asking a WTP question when the situation is viewed as WTA. This would, however, be an indication of an inappropriate use of the WTP situation and may merely reflect poor design.
7. See Spash (2000) for a detailed discussion of the issue.
8. This issue will be returned to in Chapter 7.
9. This is particularly the case with the in-depth group approach used by Clark *et al.* (2000), which meets on a number of occasions following the initial interview.
10. Lienhoop and Macmillan (2006a) demonstrate that using the Market Stall approach (discussed at length in Chapter 7) it may be possible to better investigate the perceived entitlement structure and significantly reduce protest responses.

11. The double-bounded approach can also be extended to further bounds, however, Cooper and Hanemann (1995) and Scarpa and Bateman (2000) suggest further bounds provide little efficiency gain beyond that achieved by the first follow-up.
12. Calia and Strazzera (2000) considered the comparative efficiency of single and double-bounded methods for different survey sizes and found the efficiency gains for small sample surveys to be particularly large.
13. This issue is given specific consideration in Chapter 7.
14. For example, Svedsäter (2003) found 19 respondents (66 per cent) to doubt if the policy considered would be introduced in the near future.
15. See Margolis (1982) for an economic account of the 'fair share' motivation.

6. Investigating sensitivity to scope

INTRODUCTION

The controversial nature of stated preference studies has been described in previous chapters. This chapter focuses on a specific validity issue, namely the sensitivity of valuations to the characteristics of the goods under investigation. Concerning this issue, academic debate has focused particularly on the sensitivity of welfare estimates to the scope of the goods considered. According to Carson and Mitchell (1995), scope may be defined either in terms of single argument (quantitative nesting) or across multiple arguments (categorical nesting) within a utility function. Given that respondents are not satiated in their preferences and perceive the probability of provision as constant, economic theory suggests that willingness to pay should be non-decreasing with respect to increases in the scope of a good. From the viewpoint of behavioural psychology, a measurement procedure is judged by the extent to which the responses are sensitive to relevant changes in task and insensitive to relevant changes in task (Payne *et al.*, 1992; Fischhoff *et al.* 1993). As such, scope sensitivity is a measure against which stated preferences can be judged and Kahneman *et al.* (1999) illustrate that it has been commonly observed within other experimental behavioural studies. In terms of stated preference the observation of scope sensitivity is fundamental to the applicability of the approach. If responses are insensitive to the specifics of the scenarios considered, this severely restricts the use of valuations within policy making, where it is unclear to what extent valuations can be apportioned to a specific scheme and be used within cost-benefit analysis.

Early CV applications observed scope sensitivity within same-sample sequential valuations (where a series of related goods are valued in a given sequence). These 'internal' validity tests suggested that responses were consistent with economic theory (Brookshire *et al.* 1976). Kahneman (1986) proposed an alternative split-sample ('external') validity approach in which different respondents were presented with alternate scope levels and the willingness to pay stated was then examined for evidence of scope sensitivity. Empirical findings reported by both Kahneman (1986)[1] and Kahneman and Knetsch (1992)[2] suggest that while within sample responses exhibited scope sensitivity, split-sample responses did not. These contradictory findings were explained in terms of an inadequacy with or within the same-sample test, where scope sensitivity may have been induced through respondents striving to be internally consistent with their prior valuations of a more inclusive good. As such only results of split-sample tests of scope sensitivity were viewed to be valid.

Concern over scope sensitivity was formally recognized in a report by the National Oceanic and Atmospheric Administration (NOAA) (Arrow *et al.* 1993). Following its publication, a heated empirical debate has permeated the environmental economics literature: while some studies have shown split-sample scope sensitivity (for example Carson and Mitchell, 1993; 1995; Hoevenagel, 1994, 1996; Smith and Osborne, 1996; Carson, 1997); others have not (for example Boyle *et al.* 1994; Schkade and Payne 1994); and still others show that it is possible to observe both scope sensitivity and insensitivity within the same study (for example Loomis *et al.*, 1993; Schulze *et al.*, 1998; Hammitt and Graham, 1999; Nunes and Schokkaert, 2003; Bateman *et al.*, 2004; Powe and Bateman, 2004). Reviewing this literature suggests that, although split-sample scope insensitivity can be a real problem, it is not inevitable. Given the importance of this issue, this chapter considers the contributions of mixed methodological approaches to understanding scope sensitivity.

Following this introduction, the chapter continues with a discussion of the economic expectations before considering the expectations from behavioural psychology. Evidence giving insight into the causes of scope sensitivity is then considered in terms of construal of the information provided and the motivations and strategies for valuation responses given. Before concluding the chapter, the methods for ensuring that respondents consider scenario specifics are considered.

ECONOMIC EXPECTATIONS

The economic expectations for scope sensitivity are considered through an illustration of an environmental improvement scheme, where the current level (CL) of the environmental goods and services is regarded as insufficient. A split sample CV study could then be employed to obtain values for environmental improvement of the 'whole' of the area (W) in one sub-sample and a partial or 'part' area (P) in the other. The resultant bundles of goods and services are nested, with P being a subset of the W scheme. Note that the W area fully contains and exceeds the areas covered by the P scheme. This nesting implies that the W contains at least as much of every attribute as a P.

Economic expectations regarding the relationship of values for the P and W schemes can be derived from the assumption of monotonicity.[3] However, as with most preference assumptions, monotonicity comes in two forms; weak or strong. Provision levels 'CL+P' and 'CL+W' represent the quantities of goods and services received from the combination of 'CL' and the appropriate improvement scheme. With the signs $\geq$ or $>$ referring to differences in size and quality, and $\succ$ or $\succsim$ defining differences in preference between bundles then, following the weak monotonicity assumption, if CL+W $\geq$ CL+P $\geq$ CL then CL+W $\succsim$ CL+P $\succsim$ CL. Similarly, adopting strong monotonicity, if CL+W $>$ CL+P $>$ CL then CL+W $\succ$ CL+P $\succ$ CL.

The weak monotonicity assumption only implies that marginal utility for environment improvement is non-negative. If marginal utility is zero, the individual will be satiated and additional environmental protection will not increase respondent utility. Hence the theoretical expectation for valuations made in split-samples is shown in Equation 6.1, where the equivalent loss or compensating valuation CoV(.) represents the respondents' value for the specific scheme, (that is the maximum amount of income an individual or household would be willing to give up in order that environmental quality will improve in the future, given the initial endowment provided by CL but no entitlement over additional protection).

$$\text{CoV(CL+W)} \geq \text{CoV(CL+P)} \geq \text{CoV(CL)} \tag{6.1}$$

A split-sample comparison such as in Equation (6.1) was termed by Carson and Mitchell (1995) a 'test of component sensitivity' (p. 160). Assuming strong monotonicity implies strictly positive marginal utility such that the respondent is not satiated with additional flood protection over the range considered. Replacing $\geq$ with $>$ in Equation 6.1, hence, gives a stronger version of Carson and Mitchell's (1995) split-sample component sensitivity test.

It is usual to test split-sample scope sensitivity using conventional statistical methods. However, utility theory merely suggests that respondents can order bundles of goods in terms of preference. Furthermore, the presence of substitution and/or income effects, means that we cannot simply sum values obtained for separate P schemes and compare this sum with the value accorded to predicting the whole area (Hoehn and Randall, 1989; Carson *et al.*, 1998).[4] The summation of independent valuations of 'parts' does not provide a guide as to the expected value of the 'whole'. Given this it is difficult to determine, *a priori*, the level of expected scope sensitivity between valuations of 'parts' and 'wholes'. When interpreting the magnitudes of differences observed, economists need to draw on their expectations based on other sources as to what is reasonable (Arrow *et al.*, 1993; Carson and Mitchell, 1995). The definition of what is reasonable is therefore open to debate. Further to the requirement that the difference in scope is actually perceived by the respondent, these theoretical expectations are only relevant if the goods considered are quantitatively nested, where goods differ purely in terms of quantitative attributes. However, goods are often categorically nested, differing along a number of scales (Carson and

Mitchell, 1995). Furthermore, schemes of different quantitative size may also be perceived by respondents to have differing probability of provision (Mitchell and Carson, 1989; Fischhoff *et al.*, 1993; Powe and Bateman, 2004), that is there may be variation in perceived realism across schemes. If these differences are evident then the goods are no longer simply quantitatively nested, making the predictions of economic theory in terms of sensitivity to scope less well defined. An economic expectation for the CoV measure is that the greater the perceived probability of a scheme providing the goods described (that is the higher its perceived realism), then the greater the CoV elicited.

EXPECTATIONS OF INSENSITIVITY

Using the framework developed within the last chapter, expectations of insensitivity can be determined given the cognitive and time limitations of the respondents within the interview situation; the hypothetical nature of the valuations made; and the implications of the communal nature of the scenarios and payment vehicles used.

Cognitive Ability of the Respondents

For behavioural psychologists, the observation of insensitivity to scope would be seen as an indicator of task complexity, where the expectation would be that the questionnaire process leads to preference discovery or construction rather than preference revelation. Drawing from behaviour psychology research, Payne *et al.* (1992) reports how respondents in a situation of value construction struggle to deal with the problem considered. These difficulties are likely to lead to information being neglected, some of which may be relevant to understanding the problem. Furthermore, the use of 'relatively weak' heuristics to help deal with the problem may cause difficulties, such that valuations may not reflect the scope of the scenarios considered. As such, the expectation from behavioural psychologists in the situation of preference construction would be scope insensitivity.

Fischhoff *et al.* (1993) separate problems of complexity into those relating to the scenario presentation and the response mode used. If the information presented is insufficient to gain the necessary understanding and to avoid its misconstrual, then this may cause insensitivity to scope. If the problem is adequately presented, with respondent understanding matching that of research expectations, then scope insensitivity observed may be due to the response mode used. To test this, Fischhoff *et al.* (1993) used manipulation checks to explore construal and compared the results of paired-comparisons with those of WTP estimation to test for response mode effects. Although scope insensitivity was still observed using the simpler paired-comparison approach, Fischhoff *et al.* (1993) report how reducing the cognitive task significantly alleviated the problem.

Hypothetical Nature of the Valuations

A central concern with the hypothetical nature of the valuations is the lack of incentive to take the price seriously. Hence, the incentives within the valuation exercise should be such that truthful preference revelation occurs and that the price is sufficiently large and realistic for respondents to put in the required effort. Given sufficient incentives to take their response seriously, respondents should consider the specific characteristics of the goods and services considered. In practice this is difficult to test, but using qualitative analysis it is possible to consider the motivations and strategies used; the extent to which respondents felt challenged by the price and taken it seriously; and the extent to which they have considered the scenario specifics.

Communal Nature of the Scenarios Considered and Payment Vehicles Used

Although within stated preference surveys it is up to the individuals responding to decide which issues to consider when stating their preference (Hanemann, 1994), the following of norms may lead to the use of inconsistent response strategies, making valuations highly sensitive to the particular norm followed and respondents less likely to consider the specifics of the scenarios considered. In fact the use of the economic approach assumes that respondents are able to trade-off financial gain or loss with environmental change, reducing the comparison of goods to a single metric of money. However, for respondents whose self-image is associated with green values and/or cooperation, it is difficult for them to give a non-positive response and such respondents may even object to the scenario presented. For example, in the situation of a dichotomous choice CV question, respondents in favour of the scheme may have a natural predisposition to cooperate and agree to the payment in order to avoid the guilt of non-payment and non-cooperation. Combining this hypothesis with the situation of payments viewed to be hypothetical and non-binding, respondents could try to reveal their aspirations about themselves, rather than focusing on the specifics of the scenario considered (Vatn, 2005).

A second issue relates to 'judgement by prototype', where how the choice situation is resolved is determined by the relevant properties of the situation (Kahneman *et al.*, 1999). The use of prototypes may lead to behaviour inconsistent with economic factors such as the specifics of the scenarios considered. Indeed, Vatn (2004) suggests classification or typification of a problem implies different types of rationality. Where deontic relationships exist, different approaches to reasoning could be used to those usually associated with the purchase of commodities in a market setting. For example, the social nature of environmental issues may encourage respondents to behave more as citizens than consumers, doing what is right and proper rather than following self-interest. This may be particularly the case where respondents see the situation to typify charitable giving.

CONSIDERING THE EVIDENCE

A general finding from the qualitative results is the often insufficient consideration of the specifics of the goods and services valued. For example, Schkade and Payne (1994) report a notable lack of specific discussion about the birds valued, with only 8 per cent of respondents noting the value/what the birds are worth. Likewise, Blamey (1998) did report some discussion of scenario specifics within the local responses, but reported that some individuals did not seem concerned about the specifics of the goods and services valued. The introduction to this chapter provided the judgement from an extensive literature on scope effects that, although split-sample scope insensitivity can be a real problem, it is not inevitable. This chapter does not debate this further but instead, explores the use of mixed methods and the reasons why a proportion of respondents do seem to neglect the specifics of the scenarios considered. Following a similar structure to the last chapter, the subsequent discussion is broken down into issues of construal/misconstrual of the information provided and the motivations and strategies used.

Construal of the Information

As noted in Chapter 5, Fischhoff *et al.* (1999) suggest three possibilities when misconstrual of information can occur: a misunderstanding of the original presentation; an alternative perception of the information presented; and when respondents understand and accept the information when presented but forget it when asked the stated preference question. In all three cases the respondents are answering a different question to that posed by the researcher and potentially may cause scope insensitivity.

Misunderstand the original presentation

This is perhaps the most basic of requirements and relies on the researcher's presentation skills to make the difference clear. For example, Chilton and Hutchinson (2003) presented scope differences in terms of a 10 per cent and 20 per cent expansion in the current forestry area. Whether or not respondents could actually visualize the extent of expansion was unclear, however, the results showed that respondents were at least able to see that one was greater than the other. Other methods such as mapping the areas affected have also been previously used. Clearly there is a need to test the different magnitudes within a pilot where respondents have ample opportunity to express whether the difference presented to them is meaningful or not.

The complexities of understanding the scope of scenarios considered can be illustrated through the scenario of a freshwater marshland under threat of saline flooding. Powe (2000) discusses how the saline alleviation problem in the long term would require a range of solutions including coastal defences, washlands and barrier schemes, as well as improving the river embankments, the focus of the study. Within the focus groups nearly all of the participants were aware of these potential solutions and within four of the seven groups respondents expressed difficulties considering the embankment issue in isolation from the other schemes, or rather the 'big question of how to keep the sea out'. This was despite the embankment scheme being viewed by experts to be the most pressing at the time. Given these difficulties it was perhaps not surprising that respondents also struggled to consider 'part' and 'whole' area river embankment schemes. Considering the 'part' and 'whole' river embankment scenarios in more detail, participants were asked how the areas to be protected should be chosen. This was generally seen as a very difficult question and little response was given. For example, in one group it was suggested by two participants that this was a decision for experts, a further just said 'how to choose?', and another 'any scheme to protect Broadland [the wetland area considered] is good'. In another group one participant stated that they 'can't see the point of just protecting one area', raising doubts about the 'partial' scenario considered. These examples clearly illustrate the difficulties in developing a scope test. Where respondents struggle with the different levels of scope considered, a simple solution for confused respondents is only to consider the affordability of price or cost of the scheme and leave the decision of how, and how much to protect, to experts. Such a heuristic would be insensitive to scope and is considered in more detail below.

Alternative perception

In order to develop a meaningful test of scope sensitivity it is necessary that respondents perceive the differences in quality and quantity in a similar way to that of the researcher. This problem has been demonstrated in terms of the respondent perceived likelihood. For example, despite the presentation suggesting certainty of scheme delivery, Fischhoff *et al.* (1993) and Powe and Bateman (2004) (same study as Powe (2000)) found respondents to be more optimistic about the success of more modest schemes and those considering the schemes to be realistic were willing to pay more. Essentially, this suggests that respondents were answering a different question to the one posed and in this case scope and perceived realism had opposite effects on expected WTP, weakening the scope effect. In the case of Powe and Bateman (2004) there were 41 per cent questioning the feasibility of the 'whole' scheme, whereas only between 10 and 21 per cent of respondents questioned the feasibility of the 'part' schemes. Translated in terms of scope sensitivity, this was only consistently observed when considering those seeing the schemes to be realistic. Although directly tested for in Fischhoff *et al.* (1993) and Powe and Bateman (2004), perceived divergences in presentation and perceived scheme realism were also observed within other studies not specifically considering the issue. For example, Blamey (1998) reported that respondents questioned the effectiveness of the scheme to protect the environment and Burgess *et al.* (2000) report that, due to insufficient information within the presentation, respondents were left with feelings of uncertainty as to how effective the scheme would be. Scheme effectiveness was also questioned within the studies of Chilton and Hutchinson (2003) and Powe *et al.* (2006). Hence, the perceived realism of the schemes is important when trying to understand the meaning of values elicited as well as when testing scope sensitivity and should be routinely included within survey design.

Forgetting the scope of the good

An important issue also not regularly tested, is the extent to which respondents are able to recall the scope of the goods and services being valued. In the only study known to the author where this was tested, Fischhoff *el al.* (1993) found between 21 and 39 per cent (depending on the treatment) of respondents to incorrectly recollect the approximate number of miles of river that would be improved by the scheme. This was despite a reminder. Interestingly for the issue of scope sensitivity, those respondents thinking they were getting more miles of improved river quality were also willing to pay more. Although this was a telephone survey, where it is harder for respondents to understand the issues being raised, these results were alarming. At best suggesting a shallow understanding of the scenario, where people could be considered to have merely forgotten this aspect of the problem. However, it is debateable whether they ever did grasp the scope of the goods and services being valued.

Motivations and Strategies

Although misconstrual of the information presented would evidently be a cause of insensitivity to scope, the evidence from qualitative methods suggests that the causes are more complex. As noted above, problems of scope insensitivity may be related to the complexity of the response mode used, insufficient motivation to put in the required effort and a mismatch between the individual economic approach and the communal nature of the payment and scenarios considered. Through a consideration of the motivations and strategies reported through the use of qualitative methods, this sub-section considers the relevance of these explanations.

Budget constraint

In the last chapter, evidence was presented that strongly suggested that most participants consider how much they can afford when answering stated preference questions. Although this is reassuring, it says little about scope sensitivity. Here, issues of relevance are the extent to which respondents felt tested by the price; and the extent to which responses were only reflected ability to pay and not other issues such as scope.

Considering these issues in turn, respondents need to feel sufficiently tested by the price of the goods and services valued in order for there to be sufficient incentive to consider the specifics of the scenarios considered. Within the qualitative studies there is evidence to suggest that where prices are stated, they are sometimes considered to be too small to test how much respondents are willing to pay (Schkade and Payne, 1994; Blamey, 1998; Clarke *et al.*, 2000; Powe, 2000; Powe *et al.*, 2004a; 2005). Given its more general relevance, this issue was discussed in greater detail in the last chapter.

Chilton and Hutchinson (2003) raised the issue of whether the responses were based on ability to pay but not scope. Although this issue was not new, Chilton and Hutchinson (2003) identified this motivation as an issue beyond those of charitable giving (discussed below). Within their survey, 22 respondents in a within-sample scope test stated the same WTP for both a 10 per cent and 20 per cent increase in forestry.[5] Of these 22 respondents ten clearly stated their responses reflected purely a self-imposed expense constraint, where the illustrative comments noted by Chilton and Hutchinson (2003) were 'purely due to financial constraints', 'all could afford', and 'set a household budget of £100 and wouldn't be prepared to go above'. Powe (2000) observed similar comments suggesting respondents consider their budget for the area considered rather than the specific scheme. Implicit within such a response is that the respondents are in favour of the scheme, with a comment noted by Powe (2000) perhaps illustrating this best: 'you want the scheme to go ahead and you think whether you could afford the amount stated'. This latter response relates to the dichotomous choice approach, in the case of the open-ended CV approach respondents may merely state their budget for the area considered. Together, these findings do not explain the reasons for the primary focus on the financial constraint but some insight will be given below in terms of cognitive limitations of the respondent and communal nature of the payment and goods and services valued.

Consideration of substitutes: the problem of context

An issue perhaps more directly related to scope sensitivity is the extent to which respondents consider the context to the valuations being made. Where respondents are primarily considering their self-imposed expense constraint, clearly little consideration is given to the context. In split-sample tests of scope sensitivity respondents are asked to value one good in isolation from the either, the 'whole', or the 'partial' scheme. As such, in order for scope sensitivity to be observed respondents need to consider the context, both in terms of alternative ways the problem considered can be dealt with, and the wider opportunity costs of payment. Powe (2000) provides a very good illustration of the problem of context.

Powe (2000) reports how a number of participants had a problem with considering the Broadland flood alleviation scheme in isolation from other 'good causes'. For instance, one participant (Participant D) with a bid level or price of £150 stated:

> £150 per year that's too much, it's £3 per week. I'm not paying that. If I pay that for the Broads then there is also Cumbria, the Peak District, Dartmoor ... [other National Parks].

When asked what amount would have been reasonable this participant stated: 'if it had been £10 I would have looked at it differently'. This is a positive finding suggesting the participant had considered these substitute schemes during the questionnaire interview. Indeed, this participant had said 'yes' to the principle of paying towards the scheme but 'no' to the £150 bid level. In another group, a similar problem was raised. However, on this occasion the participant had said 'yes' to the £100 bid level. An excerpt of the discussion provides a feeling for the participants' opinions in the group.

> Participant E: 'That question about money only makes sense in terms of local taxation. You can't say are you prepared to pay £10, £50 or £100 in central government taxation. I say yeah all right I don't mind to pay £100 to the Broads, but then do I want to preserve the Pine forests in Scotland and do I want to preserve something else and something else and something else ... Clearly if you have one thousand of these schemes that all cost £100, then clearly I don't want them all because I can't afford it. I can't say yes to £100 to the Norfolk Broads because I don't know that many other such schemes.

Clearly this participant had not considered this problem at the time of interview, with the problem of context only becoming apparent within the subsequent group meeting, where he had time to deliberate. Later in the same group:

> Participant E: 'Do you want to pay £100 towards the Cairngorms [a National Park in Scotland], do you want to be on a shorter waiting list at the hospital? You can't

answer it in that sort of sense. It's an impossible question, but I think it can make sense as a local taxation.'

Moderator asked: 'Is this the most important issue in your mind at the moment?'

Participant E: 'Yes, because I live here.'

Participant F: 'My biggest concern is the quality of the water coming out of my tap.'

Participant G: 'No, it forms part of my worry; what I've been involved in for several years is fighting for minimum flow rates on the rivers.'

Participant E: 'My basic problem is trying to work out how many times I have to multiply this by. If I was walking in the Cairngorms I would have answered yes, but at a limit. I can't afford to do this 100 times.'

Moderator asked: 'Did you feel that answering that question was that I am in favour of improving the environment or I'm in favour of this scheme.'

Participant E: 'Yes, that I'm in favour of improving the environment.'

Further to providing more evidence that Participant E answered the question as if the question had been 'are you willing to pay £100 towards the environment?'; it is unclear whether these participants (all willing to pay the bid levels presented in the survey) considered these substitute schemes when they answered the CV questions.

Later in the same session the moderator asked specifically how respondents would answer a similar valuation question for another area of Britain.

Participant H: 'If someone came up to me in the Cairngorms and asked me if I was willing to pay £100 per year for the upkeep of the Cairngorms I would say no.'

Participant I: 'Their own for their own.'

Participant J: 'That's right.'

Although Participant E is perhaps answering the question as if it asked 'Are you willing to pay £100 towards preserving the environment?', other participants' comments suggest they are more selective about their expenditure, focusing on their local scheme. This is fine for those respondents interviewed in the area of the scheme, but this would suggest the problem to be greater for those considering schemes not in their area.

Participants in a further group also had problems thinking about the Broadland scheme in isolation from other 'good causes'. The discussion went as follows:

Moderator asked: 'Did you think about these other schemes when you were answering the question?'

Participant K: 'I thought; well this is in isolation and if anyone else had asked me about something else the same day, what would I have said to that one? And if you were the first and I had said £100, to the tenth one I would have said I don't know, two pounds.'

This thought process is very close to the economic predictions of Hoehn (1991) and Kahneman and Knetsch (1992) where, due to order effects, the value stated depends on the placement of the good within the valuation sequence used. The discussion continued as:

> Moderator asked: 'Is this the most pressing issue to you personally?'
> Participant K: 'No, not at all. I don't even know about this area, I would like the whole of the countryside not to be spoiled ...' [tape ended].
> Participant K: 'I love trees, there should be more money spent on trees, footpaths, paths accessible by bikes. I gave this a lot more thought over the next days [following the interview].'
> Moderator asked: 'If you were interviewed elsewhere about a scheme with similar sort of amounts do you think you would be saying yes to that, or would you only be saying yes if its Broadland?'
> Participant L: 'I think if you are interviewed, wherever you go on holiday, the Cotswolds or anything, they are absolutely beautiful. If I was interviewed on holiday there I would say just the same. You've got to preserve everything in the Cotswolds that's lovely and I would probably say yes I would be willing to pay taxes towards it.'
> Participant M: 'It's like picking out a favourite charity. I mean, there are so many charities. Who do you give to? You can't give to them all. At least you could try to give to them all but if you did you would end up giving a penny to each one. So what you do is pick out two or three charities and give a pound to you, a pound to you, and a pound to you.'
> Participant K: 'And that is what you would be doing with this. What I was saying was that if this scheme goes ahead I don't mind paying. I think the actual amount doesn't mean anything.'

Following this last comment general agreement was heard.

Interpretation of the discussion in this subsection is difficult. One interpretation is that the responses suggest that if it was put to the participants that if the scheme was to go ahead they would pay their share of the costs. It would appear that at the time of the interview these respondents participating within the study by Powe (2000) considered the scheme as the only good cause they could contribute to and by implication became their favourite. This is consistent with the availability heuristic where by focusing on one environmental good, the good will become more salient and available in the respondent's mind, and subsequently will be valued higher than when it is presented in the context of other goods (Hoevenagel, 1994). An issue of more concern is that in some cases the Broadland scheme may have been symbolic for environmental protection in general, suggesting that the response given does not relate specifically to the schemes considered. Only in the case of Participant D (the first quote mentioned) was there evidence that substitutes had been considered during the actual survey observed. If respondents genuinely see CV questions as contributing to a charity this may suggest that a strategy is pursued that maximizes their moral satisfaction, in a similar way to a charitable donation.

The problem of context illustrated by Powe (2000) has also been described within other studies. For example, Clark *et al.* (2000) noted the similar difficulties of trying to value one project in isolation from the range of national problems, where there was agreement that the value of any one scheme could only be determined in relation to the wider problem. In the case of the study by Clark *et al.* (2000) participants suggested that due to the scientific and context complexities, they felt that it was not possible for them to give meaningful answers. For example, one participant stated: 'I've come to the view that I need to contribute nationally, and then have someone to even it out for me' (p. 52); and another, 'You don't know what is on the next page: would you put your hand in your pocket for this, and this, and this? You can't take one area in isolation' (p. 53). Other comments were also similar to those described within Powe (2000).[6] To a lesser extent, Svedsäter (2003) reports the issue of context in terms of wider issues such as education, health and transport.

Using choice experiments, Powe *et al.* (2005) found environmental substitutes were not discussed within the meetings, however, this was perhaps to be expected as three different types of environmental goods were implicitly considered within the choices made. Interestingly, as reported in Chapter 5, Powe *et al.* (2005) provide an illustration of what is possible in terms of understanding the context of environmental valuations. Generally, participants found the trade-off between environmental quality, service and water price relevant and most responses reflected a balance between these issues. Some participants stated that they could choose between environmental attributes and a statistically significant difference in preference was observed. However, participants generally found such choices more difficult than merely trading-off between the environment, service and water charges. In fact, some thought they had inadequate knowledge and experience in order to make valid responses and suggested that such decisions should be left to experts. These results illustrate that even when explicitly introduced within a choice experiment, some participants are unable to make the required trade-offs between alternative schemes. Hoping that such trade-offs will be made, when it is left purely to the respondent to consider context, it is likely to lead to a number of participants not doing so. As such if the scale of the scheme considered is widened, the possibility of scope insensitivity is likely. This is particularly the case when considering a local scheme with a national payment vehicle as illustrated by Clark *et al.* (2000) and Powe (2000).

Broader good – principled response

Given common neglect of context within stated preference valuations, a respondent strategy to deal with the complexity of environmental issues might be to 'do as much as they can to alleviate the problem', or as Powe (2000) reports: 'how I personally took the question was would I personally be willing to pay a little bit more in taxation then the government would undertake this work. My answer to that would be yes.' As such, respondents may consider their ability to pay and, if the particular environmental problem considered is worthwhile, state what they are willing to contribute towards solving the particular problem. However, some responses may have less meaning, where Powe (2000) reports one participant who's message was purely the principle 'preserve the Broads [area being protected], preserve the Broads'. Regardless of whether such a principled response is checked by ability to pay or not, such as response is clearly insensitive to scope for any sub-scheme considered in the area and leaves the issue of how the environment will be protected or improved and the extent of the change to the agency responsible for delivery. Although perceived by the general public to be the natural solution, in practice leaving it to the 'experts' is far from straightforward. For example, as discussed in Chapter 2, there is also likely to be debate between scientists as to the 'best' environmental solution, as well as any differences between pressure groups. More generally, these findings demonstrate how the meaning of the stated preference response may be simpler than that assumed by economic theory. For example, given the extent of principled responses within the study by Vadnjal and O'Connor (1994), had they performed a scope test the outcome would have been almost definitely insensitive to the specifics of the scenario considered.

Charity like responses

Charity like responses occur where respondents struggle with the novelty of the elicitation methods used and search for something similar or a reference point to help them decide on their response. In the previous chapter it was noted that charity like responses are related to the achievement of moral satisfaction and consist of two components; contribution and warm glow. Using the contribution explanation, respondents may still be consistent with ability to pay, but also reflect their willingness to 'do-their-bit' for environmental concerns in general, rather than the specific goods and services considered. Warm glow responses may be unrelated to ability to pay.

Evidence was provided in Chapter 5 suggesting responses tend to be of the contribution form. Rather than replicate these results, it is sufficient here to add the results of Chilton and Hutchinson (2003) who found charity like responses to be an important determinant of same-sample insensitivity to scope. Chilton and Hutchinson (2003) report that of the 22 respondents reporting the same WTP for both 10 per cent and 20 per cent increases in forestry, ten used charity like explanations, where typical comments given were: 'donation would be given irrespective of the level of planting carried out'; and 'make a donation regardless of the level of planting – have a maximum that could contribute to the cause' (p. 71). As such the evidence suggests the heuristic of 'do-their-bit' within ability to pay is common and a key reason for insensitivity to scope.

ENSURING RESPONDENTS CONSIDER SCENARIO SPECIFICS

Having provided an insight into the potential causes of scope insensitivity, approaches to encourage respondents to consider scenario specifics are considered. Qualitative methods can be used to improve survey design and analysis and gain insights into the applicability of existing methods. The potential for alleviating and controlling for scope insensitivity will depend upon the applicability of stated preference to the scenarios considered.

Improving Survey Design and Analysis

Comments raised in the discussion that followed Schkade and Payne's (1993) verbal protocol analysis illustrate the extent of the problem.[7] Schkade was asked the following question by Rosenthal from the Department of Energy:

> Your work showed that people were focusing on a lot of factors when they answered the questions, but that a lot of them were not focusing on the value of the birds themselves. Do you think that it's possible to design the instrument in such a way that they would focus on that? And, if the answer to that is 'yes', how would you have changed it so they would focus on the value of the birds themselves. (pp. 298–9).

In reply Schkade stated:

> It is our belief that the sort of percentages of these different considerations could easily be shifted around by changing the commodity, changing the payment vehicle, and other things of that kind. Whether it would be possible in principle to devise a survey where even a majority of the respondents were doing the right kind of economic trade-off, I think we're a little less optimistic about that. (p. 299)

Alternatively, if practitioners believe that respondents are fundamentally able to provide meaningful answers to stated preference questions then insensitivity to scope is seen to be due to a scenario mis-specification in survey design. Referring to bias caused by amenity mis-specification Carson and Mitchell (1995) state that: 'although these biases pose a methodological challenge to CV researchers and require careful attention in the design phase of a study, they are often avoidable if the scenario is plausible and the good is carefully described' (p. 163).

In fact Mitchell and Carson (1989) have suggested that, in the case of a poorly described good, respondents may confuse it with either a larger or smaller good (part-whole bias), or may see it as having a general symbolic meaning (symbolic bias). The respondents may also perceive the metric used by the researcher differently to that intended (metric bias), and there may be scepticism whether the good could or would be provided (probability of provision bias). These problems are consistent with the evidence above and by understanding and through using qualitative methods to enhance understanding and better testing questionnaires it is likely that these biases can be reduced. Further to improving design, the use of pre-survey qualitative methods can improve the analysis by enhancing understanding such that, for example, if respondents are more optimistic about the success of more modest schemes data can be collected on perceived realism and this can be controlled for within the modelling and, where necessary, some form of calibration can be used such that the valuations reflect actual uncertainties.

A further survey design issue relates to the possible failure of respondents to consider the context of valuations. This is a problem of the cognitive difficulty of the valuations made. Concern over the lack of consideration of substitutes was raised by the NOAA panel, which suggested a forceful reminder directly prior to the main valuation question (Arrow *et al.*, 1993). Despite this, subsequent studies by Loomis *et al.* (1993) and Neill (1995) have found substitute reminders to be ineffective. Perhaps more is required than a simple reminder. This would be likely if the question being asked is inconsistent with the cognitions of the respondent. For example, Green and Tunstall (1999) suggest that there may be a 'basic category level' at which respondents hold a preference. When asked a question at a more disaggregated level, the respondent may state they 'don't know' or give a response as if valuing the basic category. Green and Tunstall (1999) suggested that if the basic category is higher than the target good, then respondents should be first asked how much they are willing to pay for the basic category and then asked how this amount should be allocated between its sub-components. Indeed, it is common practice within a discussion on a specific topic, to ask general questions before giving the discussion the required focus, where Krueger states:

> the most common practice is to go from general to specific – that is, beginning with general overview questions that funnel into more specific questions of critical interest. Avoid hitting the participants with a specific question without first establishing the context created by more general questions. (Krueger, 1994: 67).

Cognitive psychologists have also developed approaches to help respondents answer difficult questions. For instance, Fischhoff *et al.* (1978) reports of a situation in which a fault tree was developed to work out the cause of a car failing to start. The fault tree breaks down the problem into a series of possible problem areas using a hierarchical structure. In the context of nested goods one approach is to use a top-down valuation approach in which the most inclusive good is considered first before the target component valuation. The NOAA panel rejected the use of such a top-down approach by suggesting it will forcefully avoid the problem of insensitivity to scope (Arrow *et al.*, 1993). Indeed, Carson and Mitchell (1995) suggest that when following such a top-down approach respondents will strive to be consistent with the notion that more of something desirable is worth more. Returning to the example by Fischhoff *et al.* (1978), in the case of the use of an incomplete fault tree both lay people and experts failed to appreciate what sources of trouble had been omitted. In such cases respondents use availability heuristics whereby they rate higher instances or occurrences that can be most easily brought to mind than others, where anything out of sight may be out of mind (Tversky and Kahneman, 1974). Due to the availability heuristic, where respondents have previously valued the more inclusive good, this valuation is likely to have a high weighting. If prior valuation approaches are to be used in the future more research is required in order to determine usage guidelines. If such guidelines are not available then the approach is open to manipulation, where the researcher or policy maker could design the valuation sequence in such a way that they get the answers they want.

Although having appeal, the top-down approach would not appear to be feasible. Given this, the only option with CV is to use qualitative methods to find the 'basic category level' and undertake valuations at that level. Further probing within groups will provide an insight as to how to apportion this value to disaggregated levels. At the basic category level an alternative way of dealing with the problem of context is to use choice experiments. Although there are limits as to what can be included within a choice card, Powe *et al.* (2005) has demonstrated that the method has potential at an aggregated level and can reach a stage where respondents are happy with the range of issues considered.

Applicability

Taking a position similar to that stated in Schkade's quote above, by looking at alternative scenarios with different payment vehicles and other characteristics it may be possible to minimize the number of respondents that demonstrate insensitivity to scope. The dangers of considering a local scheme with a national payment vehicle have been demonstrated. Ensuring a clearer perceived response-policy link may also assist the situation, focusing on the realities of the situation rather than suggesting a charity like situation. Supporting evidence was provided for this within Chapter 5.

For Kahneman *et al.* (1999), many of the difficulties of implementing stated preference surveys, including scope insensitivity, are due to the inappropriate use of willingness to pay. Focusing particularly on respondent ability the results presented in this chapter suggest different levels to have been demonstrated. For example:

- Some respondents are able to consider the specifics of the schemes considered, their context and provide meaningful responses.

- Others consider ability to pay and how much they are willing to pay towards solving the general problem identified, but their stated preferences have little further meaning.
- At a more superficial level, some respondents consider their ability to pay and if in favour of protecting/improving the environment, stating a preference as a charitable donation to the environment.
- Lastly, the responses of some respondents merely reflect whether they are in favour of protecting/improving the environment.

All levels have been observed using qualitative methods, but there would seem to be some credence in assuming that most people can achieve at least the second in the above list. Taking the example of Powe and Bateman (2004), despite the many difficulties in the study, once the difference in perceived realism between 'part' and 'whole' schemes had been controlled for overall scope sensitivity was observed. The question of whether greater understanding would have led to further scope sensitivity is however unclear.

Given these difficulties, Kahneman *et al.* (1999) recommend the use of attitudinal surveys that are more within the cognitive abilities of most respondents. This is in the spirit of Coombs' (1964) goal of identifying the most demanding presentation and response mode without unduly straining or influencing subjects. Fischhoff *et al.* (1993) has not only demonstrated the difficulties with construal, but also recognized that scope insensitivity may, in part, be due to the response mode used, finding the use of a simpler paired comparison approach to significantly reduce insensitivity to scope. Given the charitable typification of the scenarios and their communal nature, such a response mode may need to be implemented within a group situation such that individuals have the chance to discuss with others and have the opportunity to deliberate on the issues. The potential for using such approaches will be considered in the next two chapters.

CONCLUSIONS

This chapter has focused on a specific validity issue, namely the sensitivity of valuations to the characteristics of goods under investigation. Concerning this issue, academic debate has focused particularly on the sensitivity of welfare estimates to the scope of the goods considered. Further to the expectations of economics of non-decreasing valuation with respect to increases in scope, and of behavioural psychology that constructed preferences will tend to be insensitive to relevant changes in task, scope sensitivity is essential if stated preferences are going to play a major role within policy development.

The economic expectations of sensitivity based on micro-economic theory have been described, as well as the expectations of insensitivity based on the cognitive and time limitations of the respondents within the interview situation; the hypothetical nature of the valuations made; and the communal nature of the scenarios and payment vehicles used. Results from qualitative studies, to differing degrees, have matched these explanations. Respondents clearly struggle with the scenarios and context to the valuations; construct their own interpretation of the information provided which may differ from that of the researcher; and at times there would appear to be insufficient incentive to put in the required effort to give meaningful responses. Partly due to the communal nature of the situation, but also due to these difficulties, respondents are seen to use simplifying heuristics, particularly, 'do my bit within budget constraint'. This may be more relevant where the scenario is viewed to have similarities with donating to a charity. Although some respondents do provide responses that are sensitive to scope, it would appear inevitable that some do not. However, the extent of the problem depends on the scenario considered.

The findings presented illustrate the difficulties of performing like-for-like tests of scope sensitivity and the need for a fuller understanding of the preferences and motivations of respondents. Only after allowance is made for the complexity of perceptions, preferences and motivations, can judgements concerning the validity of given applications and stated preference methods, in general, be made. In the next two chapters the potential for further extending the use of qualitative methods to improve information on general public attitudes and preferences for environmental scenarios will be considered in detail.

NOTES

1. This result is contested in Hanemann (1994), who states that there was in fact a 50 per cent difference in values accorded to the two levels of amenity considered.
2. The questionnaires used within the Kahneman and Knetsch study were criticized by Smith (1992) as poorly designed. It was suggested that little effort had been made to consider how people relate to the goods considered and that the description of complex inclusive commodities was only brief.
3. These assumptions are given more detail in, for example, Varian (1992) and Kreps (1990).
4. Carson *et al.* (2001) demonstrate through an illustration using private goods that substitution effects are sufficient to explain the sensitivity to sequence observed in previously CV studies.
5. The use of the open-ended method enabled Chilton and Hutchinson (2003) to identify exactly the extent of scope sensitivity observed.
6. Willis *et al.* (1996) struggled with the problem of considering a few sites of special scientific interest using a national payment vehicle from the hundreds in the country. In order to deal with this problem Willis *et al.* (1996) developed an experimental elicitation method which achieved its objectives.
7. This was considering Boyle *et al.* (1994) CV study of bird deaths in waste holding ponds in the Central Flyway of the USA.

PART III

Extending the role of the group-based approach

Introduction

So far in this book, qualitative methods have been considered in a complementary manner in which they are used to; gain a better understanding of how respondents discuss and conceptualize the goods and services valued; gain a better awareness of respondents' thought processes during the transaction and motivations for their responses; and explore the public acceptability of the valuation exercise. Chapters 5 and 6 have shown the results not only lead to better survey design, but also a wealth of information that has greatly enhanced our understanding of the meaning and acceptability of stated preferences, helping within analysis and finding an appropriate role for valuations within policy decision making.

Despite the complementary role of qualitative methods, it has been demonstrated that conventional stated preference may still make insufficient allowance for the cognitive limitations of the respondents; provide insufficient incentive for respondents to put in the required effort; and make insufficient allowance for the communal nature of the scenarios and payment vehicles. Although the use of group methods can do little to correct for the hypothetical nature of stated preference, it has been suggested that cognitive and communal issues can be better dealt with by further extending the role of the group-based approach (Sagoff, 1988; Gregory *et al.*, 1993; Fischhoff, 1997; Ward, 1999; Macmillan *et al.*, 2002; McDaniels *et al.*, 2003; Philip and Macmillan, 2005; Howarth and Wilson, 2006), but the approach taken will depend on the researchers' attitude towards the ability of respondents to give meaningful answers and the relative importance of the private/social aspects of the decisions being made.

Chapter 7 considers the view that stated preference methods produce responses that are highly sensitive to changes in presentation and elicitation methods used. Faced with these difficulties, respondents should be given sufficient opportunity to construct more considered preferences. Chapter 7 explores the extent to which this can be achieved by providing a permissive and non-threatening environment for value construction as well as the opportunity to discuss the issues with friends/family and to further research any aspects where more information is required. The implications of such efforts in terms of the valuations elicited and the consequences of violating conventional survey norms are also explored. In a similar vein, where the complexities of the situation considered make it inevitable that unstable values will be elicited, the possibility of using alternative approaches, which adjust existing valuations made in more favourable circumstances, are considered.

The issues associated with the communal nature of the scenarios and payment vehicles used within stated preference are largely ignored within Chapter 7. Instead, they provide the focus of Chapter 8, where, learning from the experiences of citizens' juries, the extent to which group approaches can be used to deal with communal issues are explored.

7. Constructing better preferences?

INTRODUCTION

Of the concerns raised in Chapter 5 regarding valuation methods, this chapter focuses on the cognitive and time limitations of using stated preference. Due to the unfamiliarity of the scenarios valued and the novelty of considering environmental goods and services in terms of WTP, responses are seen to be constructed during the interview rather than being revealed from previously determined preferences. Constructed preferences are viewed within behavioural psychology literature to be insensitive to relevant changes in the scenario (such as the quantity and quality of the goods and services valued) and over sensitive to irrelevant changes (such as minor changes in the wording or ordering of questions) (Payne *et al.*, 1992; Slovic, 1995). As such it is questioned whether using conventional individual interviews, stable and meaningful values can be elicited. This chapter explores the use of group methods within presentation and providing time to reflect as a potential means of alleviating the cognitive difficulties encountered by respondents.

Using group methods as a means of presentation enables respondents to consider the perspectives of others, gives more time for deliberation, and provides a forum for the asking of questions. Furthermore, providing time to reflect following the meetings enables respondents to discuss the issues with family/friends and, should they wish, to also undertake further research. Together these modifications perhaps provide a method more conducive to value construction. However, due to an implied inevitability that personal interview stated preference methods will fail to elicit meaningful values for the scenarios considered, the approach departs from accepted survey norms and its adoption is a move away from convention. As such, this chapter will consider both the potential for the method and the implications of contravening accepted practices within both environmental valuation and behavioural psychology.

As a possible extension to the group-based presentation method, there may be a role for incorporating valuations from conventional stated preference studies within this process. Benefit transfer is usually used as a means of providing benefit estimates where there are insufficient funds for a more detailed survey. However, given concerns regarding the extent to which group methods can be used to gain a representative sample, there may be scope to feed the values from conventional and representative surveys into the decision group process. This would be particularly relevant where conventional valuations have been elicited in more favourable conditions, for example, a less complex scenario than those valued within the group-based approach and considered to be relatively more 'robust'.

This chapter begins with a discussion of opinions regarding respondent ability and the implications of different researcher attitudes in terms of the key questions to be resolved and methods adopted. This is followed by consideration of the potential consequences of adopting the constructed preference approach. A group-based approach developed by Fischhoff (1997) is then critically considered in terms of the implications of adoption and also the empirical evidence from similar post-questionnaire, market stall and multi-attribute approaches. Prior to concluding the chapter, the potential for using the results of conventionally estimated valuations within this process is considered.

RESPONDENT ABILITY

We know from the results of Chapter 5 that the presentation of standardized information within the confines of an individual interview does not always achieve the necessary 'common ground'[1] between researcher meaning and respondent understanding. Enhancing such understanding is a key motivation within survey design and the extensive efforts of practitioners to develop quality questionnaires have been previously described. For example, Lazo *et al.* (1992) and Chilton and Hutchinson (1999) have demonstrated how the standardized wording used within survey questionnaires could be improved by minimizing information on issues where there is shared understanding and by correcting for any deviance between scientific and respondent definitions. Unfortunately, Chilton and Hutchinson (1999) did not extend this analysis to explore the effectiveness of this process in terms of improving the valuation made. Testing for the effect of wording improvements, Lazo *et al.* (1992) found the stability of the valuations to be comparable to those obtained using a fuller information statement. However, as the shortened information statement was still 12 pages long, this perhaps indicates the requirements for complex goods to be able to achieve meaningful responses.

If stated preference methods are to provide policy relevant welfare estimates it is necessary to apply them to situations where there are no existing markets or transaction mechanisms through which choices are revealed. Furthermore, particularly in the case of non-use values, respondents have less than perfect information on the scenarios considered. Although the extent of preference construction within the questionnaire is likely to depend on the nature of the scenarios considered, the results presented in Chapter 5 suggest that in some situations, even after an extensive process of questionnaire design, problems of achieving shared understanding between researchers and respondents are still apparent. The implications of these findings are still debatable and ultimately how this evidence is viewed is likely to depend upon the researchers' attitude towards the ability of respondents to articulate their preferences.

Considering this issue, Fischhoff (1991) suggested three philosophies of belief in respondent ability: articulated values; basic values; and a partial perspective. The first two philosophies are seen to be at opposite ends of the spectrum of ability, with the third taking an intermediary position. Although to some extent these philosophies represent a caricature of research opinion, they are useful to illustrate important concepts and implications of belief.

The 'articulated values philosophy' is adhered to when researchers have 'enormous respect for people's ability to articulate and express values on the most diverse topics' (Fischhoff, 1991: 839). Here it is assumed by researchers that respondents have the ability to answer questions only if good enough questions can be formulated. As Fischhoff states, for practitioners of the articulated value philosophy 'the test of success is getting the question specified exactly the way that one wants and verifying that it has been so understood' (1991: 840). Fischhoff (1991) states that the concerns of those favourable to the articulated values philosophy are: inappropriate default assumptions; the possibility of question misinterpretation; difficulty in expressing values; and strategic responses. Stated preference practitioners have previously raised all of these concerns.

Viewing stated preference from the 'basic values philosophy' would mean a different focus on estimation problems. Researchers that adhere to such a perspective consider respondents to: 'lack well-differentiated values for all but the most familiar of evaluation questions, about which they have had the chance, by trial, error and rumination, to settle on stable values' (Fischhoff, 1991: 835). Indeed, Kahneman, a critic of stated preference methods, suggested: 'users of CVM often deal with people who simply do not have the kind of coherent preference order that the theory assumes – especially in domains which lack market experience' (1986: 193).

The partial values perspective is considered to be an intermediary between the two polar cases considered. If respondents are assumed to have a partial value understanding then a balance is required between being too optimistic about respondent ability to express their preferences and too pessimistic. If respondents can partially articulate their preferences then higher information content could be achieved than using the basic values assumption. However, questions that are too complex for respondents to give meaningful answers should be avoided.

All perspectives accept that there are some scenarios where prior to the survey there is an absence of well-defined preferences and in such situations respondents have to construct their preferences during the elicitation process. Stated preference practitioners generally acknowledge that respondents do not have well defined preferences for some of the goods considered prior to being interviewed (Mitchell and Carson, 1989; Hanemann, 1994). Views differ as to whether using the conventional stated preference approach respondents can construct meaningful responses. The implications of constructed preference research from a behavioural psychology perspective were discussed in Chapter 5, where in a situation of constructed preference responses are likely to be insensitive to relevant changes in the scenario and sensitive to irrelevant changes. Given the opportunity to deliberate, discuss and ask questions, which is provided using group-based methods, it is hypothesized that this will lead to more stable valuations. Whether this is true is an empirical question that will be considered below. However, before doing this it is necessary to consider the proposed approach and the implications of its adoption.

PREFERENCE CONSTRUCTION

Since Simon (1956) argued that decision making behaviour approximates more a situation of bounded rationality than maximum utility, researchers have been investigating the psychological processes through which decision problems are represented and information is considered. A clear finding from this research is that the psychological process is far more complex than suggested within utility theory and Slovic (1995) indicates that it is generally accepted among psychologists that economic theory provides, 'only a limited insight into the processes by which decisions are made' (p. 365). Neo-classical economic theory assumes people have preferences that follow a range of axioms and the preferences are revealed within transactions.

Considering the results of surveys psychologists have observed that preferences are often constructed during the elicitation process and Payne *et al.* (1992) suggest that decisions are rarely invariant to small changes in the way the questions are asked (task) or the options are presented (context). Such invariance has rarely been observed within stated preference studies. For example: ordering/sequencing effects have been empirically demonstrated which may be larger than expected by economic theory (Kahneman and Knetsch, 1992); preference reversals have been observed when comparing choice situations to valuations through CV (Irwin *et al.*, 1993); and open-ended responses tend to be systematically lower than those generated through the closed-ended CV approach (Kealy and Turner, 1993; Bateman *et al.*, 1995; Boyle *et al.*, 1996; Frykblom and Shogren, 2000).

From the constructed preference research two key challenges emerge: scenario presentation and response mode. In terms of the former, although Gregory *et al.* (1993) suggest respondents do have 'strong feelings, beliefs and values' for environmental issues, when considering the specific scenarios there is a need for a deliberative approach to preference construction. Indeed, Schkade and Payne (1994) discuss the respondent challenges of their understanding the scenario and how they feel about these often unfamiliar situations. Considering the evidence from behavioural research, Payne *et al.* (1992) suggest that sensitivity to task and context occurs particularly when respondents face such complexity and where respondents are uncertain regarding how they feel about the issues considered. Such task complexity leads to respondents struggling to deal with the problem and using 'relatively weak' heuristics (Payne *et al.*, 1992).

In terms of response mode, important lessons can be learnt from the preference reversal literature. This occurs when comparing two goods and one is preferred to the other using choice, but this preference is reversed using an alternative approach such as pricing. These results are particularly relevant to understanding invariance of responses to alternative elicitation methods. Explanations of preference reversals relate to the compatibility between the response mode and the scenario attributes (Lichtenstein and Slovic, 1973; Slovic, 1995) and the prominence of attributes within the problem description (Tversky *et al.*, 1988). In the case of the former, when using open-ended CV the focus on the pricing of the goods and services may make this attribute more prominent and some may view pricing and environmental goods as incompatible. More generally, Tversky *et al.* (1988) suggests that the heuristics used by respondents appear to differ between task where, for example, a choice situation such as in choice experiments and dichotomous choice may invoke more qualitative reasoning (such as a lexicographic strategy); whereas a pricing exercise, such as open-ended CV, may emphasize more quantitative aspects of the problem. Other findings from behavioural psychology are also important in understanding the effect of using different elicitation methods. In the case of the comparison between open-ended and dichotomous choice CV, for example, Kahneman *et al.* (1999) illustrate the importance of anchoring, where the responses within dichotomous choice CV may be biased towards the bid levels used.

Given these findings, from the behavioural psychology research perspective, preferences in complex/unfamiliar situations are seen to be constructed and be unstable, sensitive to 'task complexity, time pressure, response mode, framing, reference points, and numerous other contextual factors (Slovic, 1995: 369). Indeed, Slovic (1995) suggests that, 'without stability across equivalent descriptions and equivalent elicitation procedures, one's preferences cannot be represented as maximisation of utility' (p. 365). If it is accepted that when valuing complex and unfamiliar goods, unstable values will be elicited using conventional stated preference, this may render the conventional approach only appropriate when valuing goods and services which are not complex and unfamiliar.[2] After considering the evidence from a number of eminent researchers

Cummings *et al.* (1986), devised a set of 'reference operation conditions' (ROC) under which CV estimates may be considered to be as accurate as using revealed preference methods (p. 104). Leaving the issue of willingness to pay and willingness to accept to discussion elsewhere (Horowitz and McConnell, 2002), the remaining ROCs are central to the argument here. Namely, (a) subjects must understand and be familiar with the commodity to be valued; (b) subjects must have had (or be allowed to obtain) prior valuation and choice; and (c) there must be little uncertainty. Where these ROCs are not achieved preferences are likely to be constructed, from a constructed preference perspective the conventional approach will not produce stable values.

If the conventional approach is insufficient to generate stable values, an alternative is required. Fischhoff (1997) suggests an alternative approach in which stated preference questions are still asked, however, the process of preference construction is improved through the use of group methods within presentation and potentially time is also given for respondents to reflect and ask others after the meetings. Alternatively, Gregory *et al.* (1993) suggest the need to decompose the holistic issue by pre-developing a value structure in which respondents consider each element separately. These approaches represent a move from conventional survey norms and this may have implications in terms of the values elicited. The discussion of these implications is left to the next section. However, a common principle related to constructed preferences is recognizing that responses reflect those of well informed members of the general public in contrast to less informed members who may hold different values. In terms of metaphors, Fischhoff (1997) sees the informed approach to approximate a 'citizens' commission', where 'a representative sample of citizens is selected to learn about an issue on behalf of the electorate' (p. 1999).[3] Slovic (1995) referring to an unpublished document by MacLean (1991), discusses 'informed consent' in the context of clinical medicine. Here there is a move away from the traditional approach of delegating authority to the physician to that of expert adviser, where decisions are instead made by the informed patient. This has parallels with the citizens' commission approach where, similarly to stated preference researchers, physicians have the difficult task of presenting the medical information in such a way that it is unbiased and sufficiently informative. Indeed, in the light of the potential sensitivity of valuations to task and context, the problem of presentation comes with great responsibility.

DEVELOPING A GROUP-BASED APPROACH

Proposed Approach

It is hypothesized that group methods can alleviate the problems of scenario complexity and respondent cognitive limitations. The proposed group-based approach adopts the principle that if respondents do not have sufficient experience with the scenarios considered and expressing WTP related to environmental issues it is inevitable that they will construct their preferences within an interview situation. As such, the method of presentation used should enable respondents to be sufficiently well informed in order to develop stable valuations that are sensitive to relevant changes in the scenario and insensitive to irrelevant changes. As described in the last section, how this can be achieved will depend on researcher attitude towards the ability of respondents to articulate their preferences. Some researchers would feel that sufficient description of the issues could be achieved within the constraints of the conventional questionnaire. Evidence in Chapter 5 suggests it is possible to reach a level of understanding that is regarded by most respondents to be adequate. However, the results from testing tell a different story of possible value instability.

Rather than following a similar very detailed individual approach to Lazo *et al.* (1992) (see above), a number of researchers have expressed the view that cognitive issues can be better dealt with by further extending the role of the group-based approach (Fischhoff, 1997; Macmillan *et al.*, 2002; McDaniels *et al.*, 2003; Philip and Macmillan, 2005; Howarth and Wilson, 2006).[4] As an intermediary position between the 'articulated' and 'basic' values philosophies, Fischhoff (1997) provides a framework for the development of an approach which gives respondents every opportunity to be able to construct their preferences through group-based scenario presentation. Depending on the complexity of the issues considered, a follow-up meeting may be required, where the gap between meetings provides participants with both the opportunity to discuss the issues with friends/family and conduct any research they deem necessary.

Fischhoff (1997) developed a series of criteria for a successful citizens' commission, where the method should:

- be accessible to every citizen, such that the respondents are representative of the public interest;
- enable respondents to receive information on all issues that they deem essential;
- require consistency in response format such that values are elicited in detail with information about their meaning;
- enable respondents to understand the scope and realism or predictability of the environmental changes at stake; and

- enable respondents to deliberate, discuss the issues with others and ask questions.

A key area where these criteria depart from conventional stated preference is in terms of information provision. Conventional stated preference requires closely controlled standardized information to be provided. Using the constructed preference approach respondents are given the level of information they feel they require and are free to ask questions. This will clearly mean a loss of cross-respondent consistency, but is based on the assumption that with, 'unfamiliar topics and heterogeneous audiences, no one wording may be interpreted similarly (and appropriately) by all respondents' (Fischhoff, 1997: 201), with Strack and Schwarz (1992) suggesting conversation is required in order to ensure understanding of meaning within standardized question formats and avoid 'response effects' where respondents look for cues in the information presented and the questions asked. Group meetings not only provide more time to deliberate on the issue, but also provide a permissive, non-threatening environment that is conducive to a relaxed discussion and enables respondents to feel free to ask questions. This is very different from the social context of a personal and individual interview in which Burgess *et al.* (2000) noted respondents felt pressure to absorb the information quickly, not because of any poor interview technique but rather because the social situation was such that they did not want to waste the interviewer's time. Further to the social context of the interview, using a group approach respondents will hear different perspectives, perhaps helping them understand how others feel about the communal issues relating to the scenarios and payment vehicles used. Hence, using the group approach respondents should have a firmer understanding of the problem and responses less likely to be sensitive to small changes in the information given. The social process may also encourage them to engage more with the issues.

A further issue related to the information provided is the loss of independence. Group meetings provide a social context and in terms of the verbal comments within a group meeting, Morgan (1997) suggests neither the individual nor the group should be considered as a separable unit of analysis. What individuals say within group meetings is influenced by what others say as well as their individuality. The modelling of valuation responses from a series of groups needs to control for any group effects and the magnitude of such effects will be an important indicator when determining the effectiveness of the group approach. Although relevant for comments made within the groups, the influence of the group discussion on individual responses elicited within a questionnaire at the end of the meetings is unclear. Whilst it is expected that the use of individual value elicitation will at least help maintain some independence between individual responses, the likely group effects should be researched by comparing how information elicited from groups differs to that of individual surveys.

The evidence considered in Chapter 4 suggests that individual and group approaches provide different types and frequencies of data, where group methods may focus more on communal/general issues and individual processes more on specific issues. The issues raised within groups may reflect those shared by group members, as well as being influenced by peer pressure (dominate group members) and other group dynamics. Even if the discussion is confined only to issues that are shared by group members, the exposure of the individual to other relevant viewpoints has to be at least as broad as within an individual interview and probably much broader. However, with the focus of the discussion likely to be different to within individual interviews, this may also influence the individual responses. Related to this, Ostrom (2000) reports the results of experimental economics to show that face-to-face communication in a public good game leads to more co-operation and, given the communal nature of the scenarios and payment vehicles used, may lead to higher WTP estimates. Furthermore, the results from such experiments have also shown how people can change their mind on the basis of the attitudes of others. For example, those initially least trusting are likely to be transformed into strong co-operators by the availability of a sanctioning mechanism such that people are unable to free ride. Where issues of trust are considered within group meetings such swings of preference may occur. In terms of the availability heuristic, described above, the more communal focus of the discussion is likely to lead to participants being more likely to consider such issues when filling in the questionnaires at the end of the meetings.

Lastly, a further key difference to the conventional elicitation approach is not immediately evident from the Fischhoff (1997) criteria. As using the group approach will imply more effort to construct preferences, fewer individuals are likely to be interviewed. Fischhoff (1997) acknowledged that there would be an inevitable trade-off between sample size and response precision, suggesting the relative benefits of a movement towards the latter should be explored within future research. One potential way forward would be to consider using a similar approach to that of best practice within focus group survey design. Best practice is to undertake three to five focus groups for each segmented group, where the key principle to be achieved is saturation, such that undertaking more groups is unlikely to learn more about how people feel towards the issues considered. For free discussion to occur, it is best practice to group similar people together and the general public could be segmented in terms of age, authority/status or knowledge and experience of the issue considered.[5] Although this may provide sufficient numbers for a respectable coverage, there is however, an issue of further concern. Given the effort required to attend at least one meeting, this may also discourage all but those interested in the topic, those less well off (if a financial incentive is given) and those with sufficient time (unemployed, retired). Macmillan *et al.* (2002) suggest that improvements in representativeness could be made through the use of quota sampling, however, these difficulties are likely to remain.

Further to those listed above, Fischhoff (1997) noted two other criteria important for method development. It was suggested the method must:

- assess the quality of its measures (for example interviewing some respondents later to assess the stability of responses); and
- be able to judge whether the estimates are good enough to be used within policy making and whether alternative procedures might have fared differently.

Essentially, this indicates the importance of ensuring that stable values have been elicited and the method was fit for purpose. This implies that given the difficulties of value construction, it is important that this method should not just be generally applied without continually testing for the quality of the estimates.

Macmillan *et al.* (2002) have developed a method called 'market stall' which involves follow-up interviewing at a later date, not just to test for stability but also as a deliberate part of the process of preference construction. Providing time to reflect following the meeting enables respondents to discuss the issues with family/friends and undertake further research. As such this extends the principle that respondents should be able to have the level of information they feel they require in order to make their decisions. The effect of short periods of time (24 hours) to reflect on valuations has previously been considered by Whittington *et al.* (1992), finding responses for private goods in a developing country context to be revised by between 16 per cent and 20 per cent. For more complex public goods this may be greater. The effect of time has also been considered by Kealy *et al.* (1990), with between 16 per cent and 28 per cent of respondents changing their mind after the two week period and WTP not being significantly affected. These results are not consistent with the predictions of behavioural psychological research (Payne *et al.* 1992; Slovic, 1995).

In summary, Fischhoff (1997) suggests the use of group methods in the form of a citizens' commission, where those selected become informed about an issue on behalf of the electorate. Although presenting a number of potential benefits, the use of group methods in this way may also contravene accepted practices within both environmental valuation and behavioural psychology (Fischhoff, 1997). For economists, key concerns relate to the loss of consistency of information provided and the extent to which a representative sample can be achieved. For behavioural psychologists a concern is likely to be a movement from non-reactive responses (respondents have sufficient experience with the issue concerned and merely reveal their existing preferences) to a constructed situation where the threat of unstable responses is great. Having described the potential and drawbacks of the proposed approach, the next section considers the experience of studies which have used group methods within valuation and provide an insight into the likely success of group-based presentation.

Respondent Attitudes Towards Using Group/Individual Methods

Little is known about how respondents prefer to be consulted. Brouwer *et al.* (1999) is one of the few studies that have considered this issue.[6] Respondent preferences for consultation were explored within a post-group meeting questionnaire. Respondents were observed to like both personal interviews and group discussions, where only in one group was there a preference for group discussion over personal interview. When asked about the need to talk to others, this was expressed by the majority of participants in six of the seven groups. When asked with whom people would like to discuss these issues a majority in four of the groups suggested 'experts', a majority in two groups cited other local residents potentially affected by the scheme and only three participants felt the need to discuss the issues with other UK citizens. Although providing support for having an 'expert' within the groups' meetings, little support is given for the proposition that people feel the need to talk to their peers when constructing preferences. However, the majority of respondents in all groups did suggest that the group discussion had improved their understanding of the questionnaire and had made them feel more capable of stating their preference for the scheme considered. Given the absence of research in this area, these findings can only be viewed as preliminary and clearly more research is required.

Results from Post-questionnaire Surveys

As noted in the last subsection, the majority of focus group participants reported by Brouwer *et al.* (1999) stated that the group discussion had improved their understanding of the questionnaire and made them feel more capable of making a decision about the good being valued. Despite these findings, Brouwer *et al.* (1999) found that the majority of participants did not want to change their responses following the group meetings. These results are difficult to interpret, with the apparent improvement in knowledge having little effect on valuation responses. However, the approach with which the group effect was considered was ad hoc, with testing improved within future post-questionnaire studies.[7]

Within the studies by Powe *et al.* (2004a), Powe *et al.* (2005) and Powe *et al.* (2006) more formal testing of the effect of deliberation was undertaken. Using the post-questionnaire approach described in Chapter 4, participants were asked to complete the questionnaire prior to beginning the discussion and then, at the end of the meetings, participants were given the opportunity to revisit the questionnaire and make any changes they felt necessary using a different coloured pen. As the focus group meetings allowed the participants to deliberate and ask further questions regarding the issues, this tested the adequacy of the initial questionnaire responses to deliberation.

Using this approach, Powe *et al.* (2004a), Powe *et al.* (2005) and Powe *et al.* (2006) found between 10 per cent and 27 per cent of participants changed their responses at the end of meetings. Although this effect raises some concern, the direction of the change differed, making little difference to the overall conclusions. Hence, for the scenarios considered, allowing deliberation is not likely to have had a significant effect on WTP, but instead on the variability of responses.

When asked to revisit the questionnaire at the end of the group meetings, it was merely explained that having had the opportunity to consider the issues further some people may have changed their mind about how they feel and it was suggested that all participants look through their answers again and make any changes with a different coloured pen. As such, participants were not specifically directed to reconsider the stated preference questions. Despite their immediacy, as the valuation questions had been the focus of much attention within the discussion, participants made changes throughout the whole of the questionnaire. In the case of Powe *et al.* (2005) of the 24 participants making at least one change to the questionnaire (49 per cent) only five participants changed their valuation responses (10 per cent) with a total of six choice cards (3 per cent) being changed. This suggests that the valuation responses within this study were fairly robust to deliberation. In the case of Powe *et al.* (2006)[8] 21 respondents made a change to the questionnaire (66 per cent) and eight changed their valuation responses (25 per cent). As the questionnaire used by Powe *et al.* (2006) contained a number of attitudinal statements it was interesting to compare changes to valuation and attitudinal statements. Despite not explicitly being discussed within the meetings, 44 per cent of participants changed their responses to attitudinal statements. This suggests the valuation responses are at least as robust as attitudinal responses.

Considering the reasons for participants changing their valuation responses, within all three studies mentioned above the changes in valuations reflected the mood of the discussion with the individual groups. Although it was not always possible to decipher the precise reasons for people changing their mind, some examples provide an understanding of the issues. In the case of Powe *et al.* (2004a) two participants in one group positively changing their valuation responses through a better understanding of the purposes of the survey and an increased trust in the agency involved; whereas three participants in another group changed their mind (negatively) following a detailed discussion which revealed that there were higher priorities for the agency than the specific scheme considered. In the case of Powe *et al.* (2005), not many participants changed their mind, however, in one group two participants simultaneously changed their valuation responses once they had a better understanding of the complexities involved within the choice cards used. Within the study by Powe *et al.* (2006) there was perhaps the most notable swing of opinion observed, where five participants (in a group of eight) changed their valuation responses. These adjusted responses followed a meeting in which two dominant participants were very outspoken against the agency responsible for delivery and this was evidently the main reason for participants changing their mind. Interestingly, this group was used to test the questionnaire for one particular subgroup and the comments made were largely untypical of that sub-group within the main survey. This observation raises concerns regarding the dependence of post-group responses on the mood of the discussion. If the group presentation approach is to be adopted it is important that such dependence is the subject of future research to gain an understanding of its extent and implications.

These tests of deliberation have provided an indication of the robustness of the valuations made. In the case of Powe *et al.* (2005) the valuations were seen to be relatively stable and the use of a group presentational method is unlikely to have made much difference to the responses. In the case of Powe *et al.* (2004a), which addressed arguably the most difficult topic (biodiversity), the results suggest an inadequacy of the individual survey. The results of Powe *et al.* (2006) are unclear. Excluding the group with the two dominant outspoken participants there was only a 12 per cent change of mind following deliberation, but including that group the average change rises to 25 per cent.

However, it could be argued that given the effects of trust observed in the main survey, people would need to discuss these issues with others and perhaps with a representative of the agency responsible for provision prior to making valuations. In which case, the group approach may be advantageous. An issue not considered using the post-questionnaire approach is the extent to which the valuations elicited at the end of the group meetings are stable to further information and time for reflection. Indeed, it is recommended by Fischhoff (1997) that retesting is undertaken at a later date. Such retesting has been undertaken using the 'Market Stall' approach.

Results from the 'Market Stall' Approach

The 'Market Stall' approach to stated preference has been developed to help deal with difficulties valuing unfamiliar and complex environmental goods and services, particularly where non-use values are elicited. This approach has been widely applied to the consideration of a range of environmental issues and has also provided insights into conceptual difficulties (Macmillan *et al.* 2002; Philip and Macmillan, 2005; Macmillan *et al.* 2006; Lienhoop and Macmillan, 2006; Lienhoop and Macmillan, 2007). The 'Market Stall' approach is similar to that proposed by Fischhoff (1997). Following an initial questionnaire to gain an appreciation of participant understanding and preferences, scenario presentation is provided within a group setting, in which respondents are encouraged to ask questions and to discuss the issues. At the end of the meetings, participants complete a further questionnaire that includes a payment card.[9]

Consistent with the suggestions of Fischhoff (1997), a follow-up is made to check if people changed their mind following the meetings. This is achieved by respondents keeping a diary of thoughts during the following week and attending a second meeting a week later in which respondents can discuss the issues further, raise any further questions and repeat the valuation exercise. In the case of Philip and Macmillan (2005) a phone call was used as a replacement for the follow-up meeting.

Considering initially the results of the pioneering study by Macmillan *et al.* (2002), they found a one week gap between meetings to be beneficial for the participants. Macmillan *et al.* (2002) particularly noted that participants appreciated the opportunity to discuss the issues with other family members and to consider in more detail the budgetary implications of the project. The effect of this extra time was that 37 per cent of participants changed their mind regarding the valuations given, but with a fairly even balance between positive and negative changes.[10] This is a much greater change than observed within the post-questionnaire studies in the last sub-section and could be interpreted two ways. Either, this suggests that WTP responses are more sensitive to time than to the opportunity to deliberate or, the instability in the values elicited is due to the complex nature of the goods and services being valued. Macmillan *et al.* (2002) also undertook a conventional personal interview survey and again found markedly different valuations. This provides the additional challenge of choosing between valuations. Regression analysis showed a 'best fit' to attitudes and preferences for the final valuations using the Market Stall approach (after the second meeting)[11] and the distribution of valuations elicited during the Market Stall to better approximate expectations than that within the individual interviews. Interestingly, in terms of the issue of response dependence on the group attended, there was no statistical effect observed.[12]

Following the interesting findings of Macmillan *et al.* (2002), Macmillan *et al.* (2006) conducted a further detailed exploration of the effects of information and time on WTP. The study considered two projects, one (red kite reintroduction), which was largely unfamiliar to the general public and a second (green energy from wind power), which was much more familiar. The various treatments are too complex to be described here, but they involved both mail surveys and the Market Stall approach. The results were extremely interesting and relevant to this chapter.

Considering the effects on the valuations elicited, Macmillan *et al.* (2006) found a similar high percentage of individuals changing their mind for both the familiar and unfamiliar goods as they went through the Market Stall process (time to think, additional information and a second group meeting). However, the overall effect of this on WTP was only significant for the unfamiliar good. Although in terms of the WTP estimates this is reassuring for the familiar good, as a high percentage of people changed their mind this still indicates instability in the responses. The comparative stability of the 'Market Stall' valuations for the familiar good was further demonstrated by the deliberate introduction of negative information. Stable values are likely to be unaffected by this. Indeed, the introduction of negative information only significantly changed the values for the mail survey results, where they were least likely to be stable. These results justify the use of higher information levels and indicate that using the Market Stall approach provides more stable values than the conventional approach. Considering respondent satisfaction, the extra information provided in the two non-conventional treatments reduced the percentage of participants wishing to have more, but surprisingly even after this process some participants still felt they had insufficient information[13] (whilst others felt they had too much information). This provides support for Fischhoff's (1997) suggestion that the information requirements of respondents differ and that it should be left to the individuals to decide what, and how much, information they deem is essential. Lastly, in terms of information sources that were most useful, the standard information provided by the researchers and the opportunity to attend the group discussion scored highly. In the case of the familiar good, own prior information was also very important. As noted above by Macmillan *et al.* (2002), opportunities to discuss with family members, was much less important than access to other sources of information.

Adapting a top-down approach similar to that described in the last chapter, Philip and Macmillan (2005) were able to deal with the issue of scope and substitutes. Their approach differed to that of the personal interview by discussing a wide range of alternative schemes raised both by the facilitator and by the participants. At an aggregate level, environmental issues were compared with that of other funding such as health and education. In terms of environmental issues, four projects were considered within the same meetings, such that participants were well informed of the context of the valuations made, and the qualitative responses elicited suggest that detailed consideration was given to this aspect. This approach was clearly superior to the much more restricted top-down approaches described within Chapter 6. Although Philip and Macmillan (2005) did not formally test for scope and sequencing effects, using a structured approach to preference construction within groups, McDaniels *et al.* (2003) found the results to be largely consistent with scope sensitivity. However, there was still evidence of some participants providing identical valuations for significantly large differences in the magnitude of the goods and services valued.

Although the Market Stall has been seen to show promise as an alternative approach to preference elicitation, Lienhoop and Macmillan (2007) demonstrate that the process does not eliminate the need for good design prior to starting the Market Stall process. Lienhoop and Macmillan (2007) describe in detail how within pre-survey focus groups a series of issues were considered concerning how the general public consider the scenario considered, perceive entitlement structures and the appropriate payment vehicle. The Market Stall approach was then designed following these pre-survey focus groups to reflect these issues.

ADJUSTING FROM OTHER VALUATIONS

The evidence presented in the last section suggests that using a group-based approach can produce values that are more stable than using personal interviews. However, this is likely to depend on a range of factors such as the previous experience of the respondents; the extent to which non-use values are elicited; and the complexity and novelty of the scenario. Furthermore, concerns are raised that using the group-based approach is less likely to generate responses that are representative of the population of interest, making aggregation difficult.

Benefit transfer provides a possible means of dealing with these concerns. Conventionally used, benefit transfer is a means of providing benefit estimates where there are insufficient funds for a more detailed survey (Navrud and Ready, 2006). However, given the difficulties often encountered dealing with the complexity and novelty of the scenarios considered, the proposed approach here is in terms of adjusting from more 'robust' valuations elicited in more favourable circumstances. Such valuations could have been elicited using conventional stated preference surveys or through other means such as revealed preference or expert judgement. Where valuations are adjusted from WTP elicited using a conventional questionnaire survey, this would at least alleviate concerns that group methods are not necessarily representative.

This section will initially consider the views of Gregory *et al.* (1993) and Kahneman *et al.* (1999), who suggest a means of adjustment based on respondent attitudes from a single valuation. Practical experiences of this approach will be considered. Further experience will then be considering using more conventional valuation estimates, in terms of extrapolating from existing valuations and widening the context of a valuation study for a complex good.

Adjustment from Existing Valuations

Approach suggested by Kahneman *et al.* (1999)

Kahneman *et al.* (1999) provide a critical review of stated preference from a behavioural psychology perspective. Using illustrations mostly from their own research, Kahneman *et al.* (1999) suggest that stated preference responses are susceptible to the way questions are asked, insensitive to scope and dependent on the context within which the values are set. As these results are similar to those reported within other attitudinal surveys (Payne *et al.*, 1992; Slovic, 1995), Kahneman *et al.* (1999) suggest 'there is no stable preference order to be measured' (p. 221). It is also suggested that consumer theory does not capture the nature of people's values for environmental goods, but instead responses are more complicated and better described as reflecting attitudes rather than preferences due to the 'unavoidable consequence of basic cognitive and evaluative processes' (Kahneman *et al.,* 1999: 221). As such, although acknowledging that people do value environmental goods and are willing to pay to preserve them, WTP responses are seen as having properties of attitude, where the sign and intensity of emotion are measured on a monetary scale. Although it was suggested that WTP responses still convey useful information about reactions to policy proposals or market/referenda outcomes, as WTP responses are more susceptible to anchoring[14] and the lack of a common modulus,[15] attitude responses are seen to be superior to those from stated preference. The serious implication of this criticism is that robust valuations cannot be obtained using stated preference methods, even if survey design is improved.

Given the behavioural expectation of context dependence for constructed preferences, where the introduction of new goods and services significantly changes the valuations elicited, Kahneman *et al.* (1999) suggest that an alternative approach should be used where a scale of value is set up based on a set of hypothetical benchmark scenarios covering a broad range of issues. Scenarios would be selected on the basis of general public attitudes and expert judgement on other relevant considerations that the general public are prone to neglect (scope and duration of the problem were suggested). Having established this scale, the policy issue being considered would be compared to the agreed benchmark scenarios. This process would formalize the context upon which attitudes and valuations are seen to depend. Kahneman *et al.* (1999) suggest that measures of attitude are taken across the whole structure and, having set this structure up, a monetary value would only need to be imputed at one point within the structure and adjusted on the basis of attitudes. This formal approach to developing context is likely to be superior to the ad hoc top-down and discussion of substitute approaches described in Chapter 6.

Kahneman *et al.* (1999) do not suggest how monetary values should be elicited, but they could be estimated for less complex and more certain goods and services than the scenario considered and adjusted using the structure of attitudes. Although no practical applications are described, the approach suggested by Kahneman *et al.* (1999) is similar to that developed by Gregory *et al.* (1993) which adopts a group-based approach within its design. Both approaches are based on a partial perspective of respondent ability, where respondents are asked, 'to do what they can do well, not what they can do poorly' (p. 230), that is consider attitudes not valuations. The use of citizen groups to express their attitudes on a monetary scale as described above is rejected by Kahneman *et al.* (1999) suggesting this to be something that respondents would not do well.

Multi-attribute utility approach

Gregory *et al.* (1993) start from a position similar to that of Kahneman *et al.* (1999),[16] namely that respondents do have 'strong feelings, beliefs and values' on environmental issues (p. 179), but given the cognitive demands of stated preference methods people struggle to numerically quantify values. Regarding preference construction, Gregory *et al.* suggest that:

> if values are constructed during the elicitation process in a way that is strongly determined by context and has profound effects on the resultant evaluations, we should take a deliberate approach to value construction in a manner designed to rationalize the process. (1993: 186)

The use of group methods within this process is consistent with the suggestions of Fischhoff (1997), however, in a similar way to Kahneman *et al.* (1999), respondents are thought unable to provide stable values unless some formal process of preference construction is used. In the case of Kahneman *et al.* (1999) the focus was on context dependence. For Gregory *et al.* (1993) the focus is instead on the holistic approach to value elicitation used within stated preference, which requires the simultaneous consideration of a large number of factors. Instead, the holistic scenario is decomposed into a list of additive objectives of the problem that people value. This approach has been applied to the consideration of a range of environmental issues including wilderness preservation (McDaniels and Roessler, 1998), air quality (Kwak *et al.*, 2001) and restoration of an estuary (Gregory and Wellman, 2001).

The list of end objectives should be inclusive of all those potentially valued. Sufficient understanding of the scenario to achieve this would be gained through consultation with experts and stakeholders, where depth and range of opinion is crucial to achieving an all encompassing list of issues. For more complex issues, Gregory and Slovic (1997) suggest that the objectives may need to be structured into a hierarchical value tree, such that a single hierarchy can be developed and agreed upon.[17] This approach follows the multi-attribute utility approach (Keeney, 1980; von Winterfeldt and Edwards, 1986) and as such the structure of objectives must follow 'utility combination rules', where adding up the utilities at the bottom level of the hierarchy should produce the total utility. This rule requires independence between attributes enabling this addition to occur without double counting.

Having generated an objective list or tree, the second stage is to assess the utilities and make trade-offs between each lowest level objective. This can be undertaken within stakeholder groups. Just like the approach suggested by Kahneman *et al.* (1999) a common modulus would be used such that each of the objectives would be assessed in terms of a score, for example, from 0–100. Kwak *et al.* (2001) demonstrate how this can be undertaken sequentially from the element perceived to be the most important to the least.[18] These values could then act as weights and consistency checks would be made to check that axioms of preference are followed.

The third stage is to calculate a total value. This can be achieved through adding up the utilities, however, there may also be a need to attach monetary values. As with the suggestion of Kahneman *et al.* (1999), using the multi-attribute utility approach it is possible to estimate total value merely with the valuation of one element at one point within the objective structure and then adjusting this to the appropriate level using the utility weights. Again, this can be based on the element which is most conducive to using stated preference. This valuation also does not need to be elicited using stated preference and Gregory and Wellman (2001) used a social WTP measure (the subject of the next chapter) whereas Kwak *et al.* (2001) used a more conventional open-ended approach.

Having estimated WTP some tests of consistency are required. Ideally, this would be achieved by having at least two valuations. It is also suggested by Gregory *et al.* (1993) that sensitivity analysis should be undertaken to appreciate which elements are most important to the valuation and these should be given further consideration within subsequent stakeholder groups to check the accuracy of the utility estimates.

The multi-attribute utility approach would seem to have benefits, particularly in terms of understanding the structure of valuations and, as with the choice method, is flexible in that the results can be adjusted for subsequent changes in policies and priorities.

Considering the drawbacks of the multi-attribute utility approach, it suffers from similar problems to that of the market stall approach, where there is an emphasis on the quality of responses rather than breadth through generating a large representative sample. It is also open to similar criticism of the use of the top-down valuation approach discussed in Chapter 6, where respondents are guiding through the process of preference construction rather than relying on respondents to sufficiently consider the context of their choice as part of the process of listening to the information provided, discussing with others and reflection. Fischhoff (1997) suggests that such an approach has not been sufficiently tested for psychological effects such as sensitivity to changes in task and context.

Adjusting from Similar Valuations

This subsection considers the results of applications to adjust WTP figures from other similar valuations. The first approach considered is that of an expert panel.

This is demonstrated for an air pollution case study. Consistent with the topic of this chapter the remaining two applications use group-based approaches to consider biodiversity valuations. The first of these explores the use of group methods merely to explore the possibility of extrapolation beyond the values elicited within a more conventional study. The example used illustrates the importance of consultation before undertaking such exercises. The second example then provides an illustration of using valuations from one study to extrapolate to a similar but more complex good.

Using an expert panel

The UK Department of Health (1999) needed to estimate the economic value of improving air quality. Although it decided to adopt a WTP approach to assess the value people place on reductions in risk, that is, the value of prevention of a statistical fatality (VPF) from air pollution, there were no direct studies available that addressed this problem. Instead, a group of experts was formed to adjust from the predetermined value used by the then Department of Environment, Transport and the Regions (DETR) for the prevention of a road fatality. The expert group formed had experience in economic valuation of safety measures, environmental economics, health economics, risk analysis and the health effects of pollution. After much deliberation and consideration of the most relevant academic studies, the WTP-based values for the prevention of a road fatality were modified using factors that influence people's WTP for avoiding particular risk, namely the type of health effect (lingering or sudden), risk context (voluntary, responsibility, and so on) futurity (sooner or later), age, remaining life expectancy, attitudes to risk (younger people are less averse to risk), state of health related to quality of life, level of risk exposure, and wealth/income/socio-economic status.

Adjusting the DETR's road VPF of £847 580 (1996 prices) for risk context produced an air-pollution base-line VPF of around £2 million. This value was then modified to account for the other factors such as age, impaired health state, futurity, and so on (Table 7.1).

The Department of Health estimated that using this procedure the WTP for a small reduction in risk per death brought forward had an upper-bound of £1.4 million and a lower-bound of £32 000–£110 000 for 1 year, and £2600–£9200 for a 1 month delay in the probability of death from air pollution.

Despite the wide range for the values estimated, this illustration demonstrates how a detailed process can be used to adjust from similar values. However, as will be illustrated in the following example, it is important to consult with the general public before extrapolating from existing values.

Exploring the potential for extrapolation

As detailed in Chapter 5, Powe *et al.* (2004a) considered WTP of water customers for a scheme to enhance biodiversity on water company property. The

Table 7.1 Adjustment of air pollution VPF by supplementary factors

Factor	Calculation	VPF	Justification
Age	£2 × 0.7	£1.400	WTP >65 years 0.7 mean value of population
Reduced life expectancy	£1.4 × 1/12	£0.120	Reduction of 1 year of average life expectancy beyond retirement age
Reduced life expectancy	£1.4 × 1/12 × 1/12	£0.010	Reduction of 1 month of average life expectancy beyond retirement age
Impaired health status	£0.120 × 0.7/0.76	£0.110	Lower quality of life (QoL) than average elderly population (0.76) and with COPD with rated QoL 0.4 (std. 0.2-0.7)
Impaired health status	£0.120 × 0.2/0.76	£0.032	Lower quality of life (QoL) than average elderly population (0.76) and with COPD with rated QoL 0.4 (std. 0.2-0.7)
Risk, wealth, income, socio-economic status			No adjustment advocated
Futurity	5 yrs : 95% 10 yrs : 90% 15 yrs : 86% 20 yrs : 82% 25 yrs : 78%		Mortality occurs at some time in future after first exposure to air pollution. Thus, future risk reduction benefits are valued at current rates discounted by pure time preference rate (1%)

Source: Department of Health (1999).

Units: £ millions, 1996 prices

unconventional payment card estimated the amount respondents could forgo from a potential bill reduction to pay for the biodiversity scheme. This method of valuation was unconventional as respondents were asked to state their willingness to forgo between £0 and £5[19] and provided willingness to forgo values that were restricted to this range. A series of focus groups were then used, in part, to consider the extent to which willingness to forgo could be extrapolated beyond this range and if the responses also had meaning in terms of WTP. The following results are taken from the unpublished transcripts from this research. The results illustrate the importance of consulting the public on the feasibility of extrapolating beyond the values elicited within stated preference valuation studies.

Considering initially willingness to forgo, the facilitator within the focus groups explored participant reaction to a bill reduction of £10 rather than £5. In one group the reaction from two participants was that the amounts were not much. A distinction was also made that if it was to be a higher amount still, such as £50, then that would have an opportunity cost associated with it or as one participant put it: 'you could actually do something with that'. Alternative uses in terms of water supply and maintenance services were suggested by participants. As such, the willingness to forgo was regarded as acceptable only for small amounts. In another group, a participant suggested that there are limitations on the role of private water companies in delivering biodiversity and at higher levels of funding such schemes would be better dealt with by government. However, this issue was debated by others. With higher amounts, it was seen as more important that better information was provided to customers. Clearly these responses suggest that extrapolation beyond the small amounts within the stated preference survey is not straightforward and should not be attempted merely by extrapolation. Higher amounts may suggest different principles.

The participants of the groups were also asked how they would feel if the only way the biodiversity projects could be funded was through a bill increase. For some participants, they would be willing to pay the amounts stated. For example, one participant stated: 'It wouldn't make much difference I would still pay it, it is very small in proportion to the total bill'. Likewise, in another group one participant stated: 'What an increase specifically for this? To a degree, yes … As long as you are educated as to what it is for and informed about it'. Furthermore another suggested:

> I don't see why it is really any different for the amounts that we are talking about in this questionnaire. It doesn't matter if it is a reduction or an increase as long as the same requirements for how the money is spent are in place.

More generally, however, there was a negative response to the idea of a bill increase. There was a consensus in one group that the water company should not increase bills in order to fund biodiversity. For example, in one group a participant stated:

> I think an increase [in the water and sewage bill] would bring about a negative response. I think a decrease as long as it was fully explained would probably bring about a positive response.

Although comments were also made relating to ability to pay and whether people were willing to pay for the scheme, one participant suggested a different perspective stating:

> I am not convinced that a water company is the best biodiversity charity I could give my money to. If they said that we are going to charge you an extra pound and this goes on ... biodiversity, I might start and ask questions. When I am forgoing a reduction I think fine they saved me money, I will sort of meet them half way, they have got more efficient in doing their business'.

Agreement was heard within the group.

Similarly, in another group:

> I would be doubtful that a water company was the right place for my money to go, because they are a water provider. I would look to some of the organisations that have been mentioned so far [RSPB and other environmental charities].

More specifically, in another group there was a discussion involving five participants that there is a need to focus on the core services and three of the participants had had bad experiences with water and sewage.

In summary, the results of the valuation exercise were only applicable to customers forgoing all or part of an annual bill reduction in the range of £0 to £5. To extend these results outside this band, or to justify a bill increase, would require assumptions that may not be consistent with the findings of this survey. Although this study is unconventional, it is also illustrative in that it is important to ask respondents before extrapolating from their stated preference responses.

Extrapolating to a more complex good

Hanley *et al.* (2002)[20] were given the task by the Forestry Commission of assessing the non-use biodiversity benefits across all woodland types in Britain. This was clearly a very difficult task, particularly in terms of the complexity of issues considered and the range of values required. Given these difficulties, it was inevitable that any approach adopted would be experimental in nature. After much deliberation, as the consideration of a range of woodlands further complicates the task, it was decided that the task difficulty was such that a group-based approach was necessary. This reflects the view that under normal circumstances biodiversity is not especially suited to the conventional approach of elicitation (Spash and Hanley, 1995; Christie *et al.*, 2006). It was also decided that, in keeping with the method of Gregory *et al.* (1993) and Kahneman *et al.* (1999), a structured approach to preference construction was required. In order to ease this task further, it was decided that a benchmark figure taken from other woodland valuation studies would be used from which the group participants would adjust their valuations.

In order to estimate a valuation which can be adjusted for other types of woodland, a list of criteria was developed to aid the selection of previous valuation studies. As no one study fulfilled all the criteria, the researchers came up with a second best solution. As the study by Garrod and Willis (1997) met most of the criteria this was selected. Garrod and Willis (1997) estimated general public non-use biodiversity value for remote coniferous forests in Britain. As such this provided a valuation for one type of forestry, from which adjustment could be made to others. Given that this study was based on a large scale survey, at least some claims of representativeness in terms of the general public as a whole can be made.

Further to the difficult task of devising the information to be presented in the group meetings, it was also necessary to develop the experimental valuation approach. The approach designed consisted of the following stages:

1. Introduction to the meeting and topic, a simple explanation of biodiversity.
2. Information sheets containing the summarized details were handed out and talked through by the facilitator.
3. the group members were divided into sub-groups, each to consider a specific woodland type. Each group attached good and bad points about the woodland type they were considering to a flip chart and a spokesperson nominated to report back to the whole group.
4. The group re-convened as a whole and was asked to divide 100 tokens between all forest types. An advantage of this approach is that it reduces the cognitive task for the participants by not considering WTP. The task was very visual, with a drawn box for each forest type written on a piece of paper. The facilitator moved the tokens as instructed by the participants until an agreement (if possible) was reached.
5. The group were informed of the results of the study by Garrod and Willis (1997). As part of this exercise the average household willingness to pay for remote coniferous forests was written into one of the boxes. Following a detailed explanation of where this value had come from and subsequently gaining agreement that the group participants were willing to pay this amount, they were given the task of trying to put WTP figures in the other boxes.
6. Throughout the exercise participants were told that they would be given the opportunity to state their own, rather than group opinion at the end of the meetings. The last step was the completion of a questionnaire that repeated the group task.

To ease the already difficult task, each group only considered either a range of upland woodland types or lowland woodland types. This kept the number of sub-groups to a manageable level. A total of seven groups were undertaken (four in England, two in Scotland and one in Wales), with there being eight participants in most of the groups.

The method generally worked well, with the following observations about the process:

- Despite being explicitly instructed to focus purely on non-use biodiversity issues participants found this difficult and were sometimes observed to also consider recreational benefits.
- Participants found it easy to use tokens to assign relative weights between different types of woodland. Most groups managed to reach agreement, which may have been helped by the knowledge that participants would get the opportunity to fill in a confidential questionnaire individually at the end of the meeting. The group and individual scores were very similar.
- Difficulties were encountered convincing the participants of the importance of the WTP exercise. However, in all but one group, all participants were eventually convinced of the need for the valuation. Difficulties were also encountered explaining the procedure used in the Garrod and Willis (1997) study and the accuracy of the valuations elicited were questioned. This demonstrates the difficulties of introducing a valuation from which to adjust, where agreement needs to be gained for its use to be effective. Similar WTP estimates were elicited within the group and individual responses.
- No simple linear transformation between tokens and WTP amounts was observed, suggesting that these tasks may be viewed as different and requiring the use of different heuristics.

Although providing some useful 'ball park' estimates of WTP, the process must be considered as experimental. It is useful in understanding what can be achieved using a more structured approach to value construction. The use of the tokens was found to be a cognitively simple task, but not necessarily consistent with one where real money is involved. This is clearly an area for future research as the approaches of Gregory *et al.* (1993) and Kahneman *et al.* (1999) depend on the input of just one (or perhaps two to check consistency) values. If a process apportioning tokens or utility is different from apportioning money, then these differences need to be understood. However, the use of the fixed amounts taken from the study by Garrod and Willis (1997) further complicated the process and it was clear there was a lack of understanding of the valuation process used. For reasons of time spent on the task, level of understanding and cognitive simplicity of the task, it was suggested by Hanley *et al.* (2002) that the results from the tokens are likely to be more accurate than the WTP exercise. Clearly, given these weights, if an overall WTP for non-use biodiversity of woodland in Britain could have been elicited, it would be a simple task to apportion WTP by woodland type.

Interestingly, an expert group was also formed which followed the same process of the general public groups. As expected, experts had little difficulty with the task of understanding the information presented, however, having this wider knowledge, they found it difficult not to struggle with the inevitable simplifications that had been made within the design of the information sheets. Experts apportioned their tokens and WTP differently to that of the general public. Based on their knowledge of the rarity of the biodiversity considered, this issue was given particular consideration. Interestingly, however, they found it harder than the general public to reach a consensus. This illustrates the inherent uncertainties and lack of agreement between experts on the issues considered. These complications need to be kept from the general public, as they would further complicate what was already a very difficult task.

DISCUSSION AND CONCLUSION

Describing complex scenarios using words and other media within the limited period of time provided within a personal interview is a challenge. During this process respondents construct their preferences and, consistent with the findings of other behavioural psychology research, the values elicited have a tendency to be unstable and sensitive to minor changes in the scenario presentation. This is particularly the case whenever the scenarios are complex and non-use values are being elicited. In the presence of evidence suggesting an inadequacy of the conventional interview situation, this chapter has explored the potential for using group-based approaches and time within value construction. Such groups provide a permissive, non-threatening environment that is conducive to a relaxed discussion. They also enable the participants themselves to decide what they need to know and to ask for the additional information they require. Giving time to reflect following the meetings also allows participants to search for further information and talk to friends and family.

Considering the evidence, in the case of unfamiliar scenarios the use of a group method would seem to produce different and in most cases better quality valuations to those elicited through individual interviews. Particularly evident is that respondent understanding of the issues is superior to that achieved within a conventional individual interview. A further benefit is that using the group approach enables the analysis of the discussion in which participants provide insights into the meaning of the valuations expressed. Providing follow-up meetings or interviews after a time interval would also appear to improve respondent understanding and the valuations elicited.

As the use of group methods breaks conventional survey norms, its adoption does not come without methodological implications. Specifically, the use of the group approach no longer maintains cross-respondent consistency in the information provided; may lead to a different focus on communal issues; and there are questions regarding the ability of such an approach to provide a representative sample of the population of interest. Although using the group approach does appear to improve valuations, it is not a panacea and in the case of highly complex and unfamiliar goods and services these improvements may still be insufficient to generate stable responses. However, the approach, although imperfect, does lead to an enhanced understanding of general public attitudes and preferences for the goods and services provided. These data are likely to be informative within the decision making process for the schemes considered.

Where the complexities of the situation considered make it inevitable that unstable values will be elicited, the possibility of using alternative approaches which adjust existing valuations made in more favourable circumstances was considered. The potential for achieving this was considered in terms of using group-based methods with experts or the general public. Examples are provided that suggest such an approach is by no means simple and should only be undertaken with care. The importance of this is illustrated using the example where group members were asked if valuations expressed could be extrapolated beyond the range originally elicited. The results showed that for some people the principles by which the initial valuations were elicited would change if extrapolated and strong opinions were expressed against extending the use of the values outside their original range and context. In a further example, the use of an apportioning task was found to be cognitively simple for the participants and the results useful as this would enable the WTP for a specific survey to be widened from one type of woodland to a much wider range of woodland types. However, when the apportioning task was extended to WTP it would appear the principles changed. Comparing the apportioning of tokens to WTP respondents may use different heuristics and these results illustrated the need for further research to consider which approach is the most appropriate.

In conclusion, if it is inevitable that different approaches to elicitation will emphasize different aspects of the problem and generate unstable valuations, responses can only be evaluated in the context with which they were elicited. If such responses are still useful within policy making, there is a need to use an approach to preference construction which is generally acceptable to the general public. It is argued that this should be decided through using groups of informed citizens, who have experienced a range of elicitation methods as well as hearing the relative arguments of the alternative methods. This further illustrates that when dealing with preferences for environmental goods and services, nothing is simple.

NOTES

1. This was the term used by Strack and Schwarz (1992).
2. Interestingly, there is empirical evidence suggesting invariance to task being upheld when valuing private goods with which respondents have experience (Kealy and Turner, 1993; Boyle *et al.*, 1996).
3. Although also acknowledging that market and referendum metaphors may be appropriate, a 'citizens' commission' was seen by Fischhoff (1997) to be particularly relevant as it embraces the informed nature of constructed preference.
4. For Howarth and Wilson (2006), as others, the key motivation for this suggestion is the elicitation of the social WTP, which is based on consensus rather than individual responses. A key motivation of Howarth and Wilson (2006) for taking this approach is to deal with the communal nature of the scenarios and payment vehicles used. Chapter 8 is dedicated to the consideration of these communal issues. The topic for discussion within this chapter is dealing with difficulties of gaining sufficient understanding.
5. See Chapter 4 for a more detailed discussion of this issue.
6. See also Macmillan *et al.* (2006) discussed below.
7. The author was the facilitator for the group meetings within Brouwer *et al.* (1999).
8. The group questionnaires for Powe *et al.* (2004a) are no longer available.
9. A payment card consists of a list of values covering the range of monetary amounts that the vast majority of the general public would be WTP. In the case of the Market Stall approach somewhere in the region of eight monetary amounts were listed and the participants asked to state, for each amount, the degree of certainty that they attached. The choice options for each monetary amount were: definitely pay; probably pay; not sure; probably not pay; and definitely not pay. WTP estimates can then be made using the approach of Welsh and Poe (1998).
10. Philip and Macmillan (2005) and Lienhoop and Macmillan (2006a) report a similar percentage of people changing their mind and a balanced pattern of changes in both positive and negative directions.
11. Lienhoop and Macmillan (2006b) also observed the best fitting models to be those undertaken using the Market Stall approach.
12. This was the only mention of testing for group dependence in the Market Stall studies and there was no further qualitative research into the more subtle effects of group dependence that may have occurred.
13. Lienhoop and Macmillan (2006b) found results to these as well as the observation that in a conventional personal interview situation 13 per cent of respondents found the process too demanding, whereas in the Market Stall approach none of the respondents felt this way.
14. Such anchoring is a particular problem within the dichotomous choice approach to CV.
15. Attitude responses are consistently elicited within a fixed scale, whereas WTP responses are limited only by perceived ability to pay.
16. See also Gregory and Slovic (1997).
17. For an actual example see Gregory and Wellman (2001) who provide a fundamental (ends) objectives and fundamental (means) objectives figure that was used within their study. The fundamental (ends) objectives clearly explain the potential outcomes of the project.
18. Kwak *et al.* (2001) provide excellent appendices where it can be clearly seen how this process works.
19. This range was chosen for its immediate policy relevance, however extrapolations beyond this range were also considered in order to investigate the possibility of extending the scheme in the future.
20. Full details of the exercise are provided by Hanley *et al.* (2002). This includes the information and questionnaire used as well as a detailed analysis of the results.

8. Communal nature of stated preference

INTRODUCTION

When considering environmental issues, it is hoped that governments will make decisions in the interests of society. As the general public are often required to pay for environmental schemes through taxation, it would seem appropriate that they are asked for their WTP. When stating their WTP, the general public are unlikely only to consider the personal financial consequences of the outcomes but also compare the options to their held values (a belief that a specific mode of conduct or end state should occur) and personal norms (such as 'I should do my bit to help the environment'). Although such concerns are also relevant to goods and services for which there is a market, due to the communal nature of the scenarios and payment vehicles used, norms and values are more important when considering public goods. Such norms and values may not be conducive to measurement in monetary terms.

Largely ignoring the issues associated with the communal nature of the scenarios and payment vehicles, Chapter 7 focused on the psychological effects caused by task complexity. Instead, such issues provide the focus of this chapter, where the extent to which group approaches can be used to deal with communal issues are explored as well as the possibility of estimating social rather than individual WTP. In terms of the structure of this chapter, how norms and values complicate the stated preference methods are initially considered in conjunction with issues of social equity and aggregation. The citizens' jury approach to communal consultation is then detailed, before exploring the extent to which the lessons learned from its application can help provide better estimates of WTP. This discussion also considers the relative merits of estimating social WTP, where a social consensus is reached regarding the amount society is willing to pay.

COMMUNAL SCENARIOS AND PAYMENT VEHICLES

Narrowness of the Valuation Approach

The results of Ostrom (2000), described in more detail in Chapter 5, suggest an ineffectiveness of microeconomic theory in explaining behaviour concerning collective goods. Her review of experimental economics suggests a plurality of values to be held, where some actors follow self-interest and behave in a manner consistent with economic theory, whereas others also follow norms of trust and reciprocity. Individuals asked to participate in a communal project, such as those considered within stated preference studies, are likely to have a predisposition to follow these norms and agree to the payment. Failing to follow such norms, may lead to feelings of guilt, for example. The results presented by Ostrom (2000) also showed that where individuals felt they could trust other players, they were much more likely to behave in a cooperative manner. The results from qualitative studies presented in Chapter 5 have demonstrated how trust, or lack of trust in the organization responsible for receipt of payment and delivery, may significantly effect cooperation within the scenarios considered as well as how much respondents are willing to pay.

More generally, it is argued that ecological economics should not restrict itself to the neoclassical model of human behaviour (Elster, 1983). Indeed, Blamey (1998) adapted Schwartz's norm-activation model (Schwartz, 1977), which he saw as appropriate for considering cooperative behaviour where there are non-use values. Blamey (1998) also extended Schwartz's model to take account of the communal nature of the public goods and services being valued. As demonstrated in Chapter 5, the communal nature of many of the scenarios considered within stated preference studies are likely to make issues such as trust in the organization more prominent than within private transactions.

Although within stated preference surveys it is up to the individuals responding which issues influence their choices made (Hanemann, 1994), the following of norms may lead to the use of inconsistent response strategies, making valuations highly sensitive to the particular norm followed and making it more likely that respondents will be influenced by the social situation of an interview. Vatn (2004) questions the appropriateness of the neo-classical economic approach suggesting that where deontic relationships exist, different approaches to reasoning would be used rather than those associated with the purchase of commodities in a market setting. For Sagoff (1988) reducing values to WTP should be avoided, stating:

> Private and public preferences belong to different logical categories. Public 'preferences' involve not desires or wants but opinions and views. They state what a person believes is best or right for the community or group as a whole … an analyst who asks how much citizens would pay to satisfy opinions that they advocate through political association commits a category mistake … it is the cogency of the arguments, not how much partisans are willing to pay … that offers a credible basis for public policy. (pp. 94–5)

Furthermore:

> we are not consumers simply bent on satisfying every subjective preference. We insist on our role as citizens as well. And, as citizens, we adopt a model of government and a vision of political life that allow us to posit collective values and give effect to our national conscience and common will (p. 97)

Given the social nature of the issues considered, respondents could try to reveal their aspirations about themselves, rather than focusing on the specifics of the scenario considered (Vatn, 2005). If there are ethical principles or feelings of moral obligation involved in the scenario considered, some individuals may object to being asked to trade-off these principles against money. Further to complicating the interpretation of WTP responses, where these problems are severe, this also indicates that a more social and inclusive approach to value construction would be needed within the consideration of communal scenarios.

Equity and Aggregation

Using environmental valuation to elicit individual WTP, which can be subsequently aggregated within cost-benefit analysis, has been much criticized in terms of issues of equity (Jacobs, 1997; Wilson and Howarth, 2002). This objection is widely held, particularly the implicit assumption of cost-benefit analysis that the existing income distribution is optimal (Pearce, 1983). This assumes equality in terms of marginal utility of income, such that £1 creates the same additional utility regardless of whether people are rich or poor. Through the use of weighting, cost-benefit analysis could be adjusted, however, this would need to be undertaken based on some form of political consensus and is controversial. This is particularly the case as, within stated preference surveys, individuals are free to consider what they wish, and some respondents may have considered equity as part of this decision process. The equity implications could be made more explicit through the use of a group-based approach and be discussed with others prior to completing the relevant stated preference questions. Alternatively, attempts could be made to reach a social consensus within a 'representative' group.

Based on the principles set out by Habermas (1984) and Dryzek (1990, 2000) in the context of environment valuation, Wilson and Howarth (2002) and Howarth and Wilson (2006) suggest that valuation of environmental goods should not result from the aggregation of separately measured individual preferences, but from a process of free and open debate within which participants can go beyond private self-interest to consider wider ranging issues such as equity. For Wilson and Howarth (2002) equity needs to be considered in a situation within which each person is fairly represented, such that issues of intergenerational equity as well as how payment should be made and by whom are considered. It is argued by Wilson and Howarth (2002) that when considering ecosystem goods and services that are seen to be rich in equity issues, the 'most appropriate value articulation institution will be one that most closely mirrors Rawls's "original position" – a procedurally based public forum in which people are brought together to debate before making value judgements' (p. 434). In practice Rawls's 'original position' is not achievable as it requires a situation in which people:

> do not know their place in society, their class position, or social status, their place in the distribution of natural assets and abilities, their deeper aims and interests, or their particular psychological makeup … no one is advantaged or disadvantaged by natural chance or social contingencies. (Rawls, 1974: 141).

Although recognizing the need to consider equity within environmental decisions and break down barriers of social and economic position this will be very difficult in a public forum. A fair process may well lead to a fair outcome, but it is questionable to what extent a fair process can be developed. One way for the fairness of the situation to be maintained is through preferences being stated confidentially and individually.

Howarth and Wilson (2006) compared the conventional cost-benefit approach to aggregation to that of a consent-based approach,[1] in which small groups of the general public are convened to make recommendations on an environmental issue, where there will be a financial cost to the tax payer if the scheme is implemented. Considering the issues theoretically, Howarth and Wilson (2006) found that unless the benefit from the public good were 'perfectly aligned with the existing tax system' (p. 13), which is unlikely, the consensus-based approach would generate a different outcome than that generated through cost-benefit analysis. The difference between the two approaches is due to the consideration of equity within the consensus-based approach.

Community Involvement

The use of stated preference methods can be regarded as more of a passive rather than an active approach to participation (Blamey *et al.*, 2000), where the completion of a structured questionnaire and statement of WTP provides little in the way of involvement. Kenyon *et al.* (2001) criticize stated preference on sustainability grounds for not giving the community the central role within participation as suggested by the United Nations Conference on Environment and Development (United Nations, 1993). It was argued by Kenyon *et al.* (2001) that the citizens' jury, described below, better achieves this objective.

BALANCING CONSUMER AND CITIZEN PREFERENCES

Environmental issues and funding inevitably are both of a private and public nature. Using stated preference approaches emphasizes more private or consumer issues, whereas, developing a consent-based social approach is more likely to focus on the citizen aspects. Given the inevitable dual nature of the situation, the choice of consultation method will affect the results obtained. Indeed, the preference reversal literature has demonstrated that adopting a particular form of elicitation will not provide unique responses (Lichtenstein and Slovic, 1973; Tversky *et al.,* 1988). Jacobs (1997) suggests there is no neutral approach to elicitation in terms of theory and suggests decision makers and practitioners need to ask what kind of values they wish people to articulate, and then design an elicitation method to encourage the appropriate values. A key aspect of this judgement is likely to be the relative importance of private and public issues to the scenario considered.[2] The work of Frey (1997) suggests that there may be implications associated with elicitation method choice, where the focus on monetary values may 'crowd out' civic virtues.

Getting the balance right between consumer and citizen preferences will be difficult and ultimately the elicitation methods adopted will tend to favour one over the other. Since the early days of stated preference there has been a definite movement away from considering responses in the context of market situations towards one of a referendum, which is seen as more appropriate given the public nature of the goods considered. Indeed, a recent direct comparison with an actual binding referendum provided very similar results (Johnston, 2006). However, the further away from the market the method goes the harder it is for the estimated valuations to be included within cost-benefit analysis. If respondents have considered costs and/or shown empathy for others there may be issues of double counting. Critics of stated preference methods have suggested a further movement away from the consumer more towards the citizen (Sagoff, 1988; Jacobs, 1997; Wilson and Howarth, 2002; Vatn, 2005; Howarth and Wilson, 2006). For example, Jacobs (1997) argues that:

> valued parts of the natural world are widely regarded as goods to society, over and above the benefits they provide to the individual. Society is better for having them, even if the number of people who privately benefit from them is very small. (p. 212)

Given the focus of stated preference research on private benefits, it could be argued that individual responses elicited using valuation methods would underestimate existence value. Although in practice using the stated preference method sizeable existence values have been estimated, this might suggest the inappropriate use of individual monetary values elicited in social isolation instead of the expression of normative reasoning within a group debate.

Blamey *et al.* (1995) considered the citizen/consumer debate in the context of a dichotomous choice CV study of forest preservation benefits, where the relative importance of citizen-like attitudes were assessed by their explanatory power within the WTP models estimated. Blamey *et al.* (1995) devised two citizen type questions: the first relating to a general judgement as to whether more should be spent on the environment or the economy; and the second whether the government should spend more or less through taxation on the environment. With the dichotomous choice response as the dependent variable, the models estimated suggested that these citizen type variables were significant and provided more explanatory power than the bid level and the income of the respondent. Indeed, the income variable was not significant in the final pooled estimated model. This evidence was seen by Blamey *et al.* (1995) as suggesting that respondents were expressing social or political judgements rather than specific preferences over consumption bundles. As such, judgements tended not to relate specifically to the scenario considered, but instead reflected principles.

In the case of a controversial public goods policy, Blamey (1996) suggests that citizen values are likely to have been discussed widely in the media and citizen preferences are likely to be better developed prior to the survey than consumer preferences, which may not have been constructed at all. Blamey also suggests that although the dichotomous choice approach is likely to be more acceptable in these circumstances, due to its similarity with a referendum situation, it is also likely to encourage more citizen type preferences which are inconsistent with economic theory. If well informed citizen preferences are more policy relevant, then Jacobs (1997) recommends adoption of more deliberative approaches in which participants are encouraged to state citizen rather than private economic values.

Considering these ideas in the context of the evidence presented in Chapter 5, objections to make trade-offs between income and environmental goods and services; principled reasons to broader issues; and WTP a fair-share of the cost all indicate more citizen type preferences than individual. Problems reflecting the choice of payment vehicle also illustrate the complications of the communal nature of the scenarios considered.

A common criticism of the use of stated preference is that respondents will refuse to trade-off environmental concerns against personal income. The study by Vadnjal and O'Connor (1994) best illustrates this with a scenario within which responses are dominated by principles of what is right and proper rather than individual WTP. Other studies by Stevens *et al.* (1991) and Spash and Hanley (1995) have also illustrated that objections may occur when presented with trade-offs between income and the environment. The results from Chapter 5 were mixed. For example, Svedsäter (2003) found 45 per cent of respondents to claim that the environment is not a monetary issue and that policy making should not be based on 'private economic decision making' (p. 132). However, these studies were far from universal and may depend on the context of the surveys undertaken. For example, when considering environmental issues in the context of water supply or flood alleviation such objections were not common (Powe, 2000; Powe *et al.*, 2004a; 2005). On balance, considering the evidence, objections to the payment vehicle are perhaps more likely (Cameron, 1997; Blamey, 1998; Powe, 2000; Powe *et al.* 2006),[3] which is also illustrative of the problems of the communal scenarios.

Although the magnitude of the problem varied between studies, Chapter 5 reports that Schkade and Payne (1994), Blamey (1998), Cameron (1997), Powe (2000), Svedsäter (2003) and Powe *et al.* (2004a) all provide evidence that respondents see their WTP as symbolic for a contribution towards solving environmental problems in general. This can be thought of as a principled response to do perhaps 'as much as they can within their budget' towards protecting the environment, rather than considering the specifics of the individual scenario under consideration. Such a response is inconsistent with the individual stated preference approach, but whether using a more social approach would achieve a greater understanding is unclear.

Perhaps more damning is the inclination to use the 'fair share' heuristic. This is indicative of a 'category mistake' as people request the cost of the schemes and wish to pay their fair share rather than maximum willingness to pay (Schkade and Payne, 1994; Bohara *et al.*, 1998; Blamey, 1998; Powe, 2000; Svedsäter, 2003). Clearly the fair share of the cost will depend on the payment vehicle used, but determining what a fair share is might be best achieved within a group where issues of equity and fairness become the focus of discussion (Sagoff, 1988; Wilson and Howarth, 2002). Jacobs (1997) suggests that ethics are a matter of argument, which requires public discussion.

CITIZENS' JURIES[4]

Whichever method of public consultation is used, some means of aggregation is required such that the overall judgement of the general public can be considered. Environmental valuation can feed into decision making either through the summation of individual preferences to be used within cost-benefit analysis or by using some decision rule with referendum type method, for example a price that 50 per cent or 75 per cent of the sample are willing to pay (Hanemann, 1994). An alternative is to use a consent-based approach in which individuals are brought together to come to a consensus through discussion. Howarth and Wilson (2006) suggest the legitimacy of such an approach rests on the consent of the general public, where their analysis suggests that 'an operational conception of the public interest may be derived from the democratic principle that social legitimacy rests in the (hypothetical or actual) consent of the governed' (p. 3). As with the referendum approach, consent-based deliberative methods enable pluralistic values (including equity) to be considered within the outcomes, with aggregation of preferences reached either through consensus or some form of voting mechanism. It is argued by Howarth and Wilson (2006) that through improving participant understanding of the views of others, this may increase the likelihood that consensus will be reached. Perhaps the most popular form of consent-based method is the citizens' jury, which was introduced in Chapter 2.

Citizens' juries consist of a small group of individuals, usually somewhere between 10 and 25 jurors depending on the nature of the scenario considered and the objectives of the exercise. Unlike other group methods, where it is common to select groups based on similar experiences in order to best facilitate discussion, citizens' juries can require a 'representative' group of citizens to be assembled. The problem achieving representativeness is that, whereas the number of societal characteristics is large, the number of participants is small. As such, statistical representativeness is not really achievable. At best, juries can be recruited to follow the characteristics of the relevant population, but Aldred and Jacobs (2000) suggest such claims are 'dubious for such a small sample' (p. 225). Dryzek (2000) replies to such criticism by restating the argument in terms of representation of discourses rather than individuals, where the process is seen as representative if participants are able to deliberate on all possible discourses. Brown *et al.* (1995) suggests the problem of statistical representation can be alleviated through using a panel of stakeholder representatives to agree on the participants.

With regard to recruitment, Brown *et al.* (1995) makes the useful distinction between stakeholder negotiation and a citizens' jury, in which the former should consist of representatives of different groups and the latter general public judgement. In the case of the study by Aldred and Jacobs (2000) a different approach to recruitment was required in order to ensure that all the relevant interests and perspectives were voiced. The case study considered the conversion of agricultural land into wetland in Cambridgeshire, in the East of England. As such, there was the need to voice the interests of farmers in general but not those whose actual land would be considered. Three people involved in farming were included within the 16 participants.

Within citizens' juries individuals are asked to couch responses in terms of the benefits to society as a whole; acting as 'agents of society' (Brown *et al.*, 2005: 252); or as a 'community-wide perspective' (Ward, 1999: 94). Despite being instructed to consider society as a whole, in practice it is likely that jurors will also consider their personal preferences. Within a real jury there is no incentive to do so, as participants are not affected individually by the outcome. This is not the case within an environmental citizens' jury. There may be issues of right and wrong associated within the outcomes, however, given the pluralistic nature of the preferences, jurors will usually be affected personally through the taxation they pay and the benefits they receive from the environmental assets considered. Consequently, the jury metaphor is imperfect, just as the referendum or market used within stated preference. Brown *et al.* (1995) acknowledges this concern and suggests this problem can be alleviated by excluding people from the juries with 'compelling personal interest' (p. 256).

Further comparing stated preference and citizens' juries, it could be argued that, although the participants of a citizens' jury are likely to be statistically less representative of the population than using stated preference methods, they are likely to better represent the views of the participants involved in the process (Sagoff, 1998; Aldred and Jacobs, 2000). Depending on the complexity of the topic area, the meetings usually take between two and five days during which the jurors can call 'witnesses' who present additional information to the panel of jurors and are able to cross-examine the witnesses.[5] As such, the jury is given a degree of power to; define what information it requires; consider the perspectives of others; time to deliberate; and be able to become informed as to the issues considered. Given these opportunities, citizens' juries could be regarded as the most detailed approach to preference construction available.

Within their experimental study, Aldred and Jacobs (2000) report promising results in terms of preference construction. They found participants to take the process seriously and actively participate, with participant interest levels rising within the later days of the juries. Although Aldred and Jacobs (2000) report some anecdotes suggesting that a detailed understanding was achieved by the jurors, this was not formally tested. Following the jury process a focus group of some of the participants was undertaken to explore reactions to the jury. Favourable responses were made regarding the process. Interestingly, Aldred and Jacobs (2000) noted that the participants assessed the experts based on 'sincerity, commitment to their argument, communications skills and credibility of the organisation with which they were a member' (p. 229). This illustrates the non-neutrality of the process, where participants can be influenced by the quality of the presentations by the relevant interest groups. Well-organized and resourced interest groups are likely to make better presentations. In addition, there is also concern that the verdict will be affected by 'unusually articulate or passionate members of the group' (Brown *et al.*, 1995). As was noted in Chapter 7, group-based approaches are susceptible to influence by dominant members. Indeed, Powe *et al.* (2006) demonstrated this clearly for one group in which two participants were very outspoken against the agency responsible for delivery and this caused a large swing of opinion between pre- and post-focus group responses. The post-group preferences of this group of eight people were largely untypical of the main survey results. Although skilled facilitators can alleviate this problem, the example illustrates the need to repeat the exercise in order to corroborate the findings should this problem occur within a citizens' jury. Given the infeasibility of asking for such additional funding from the commissioning organization, perhaps two groups should become standard practice.[6]

With regard to the stability of values, Aldred and Jacobs (2000) report that one participant suggested in the post-jury focus group that his views had altered substantially since the jury and that he now disagrees with the findings of the process. This illustrates that even with what may be regarded to be the 'best' approach to preference construction, stability of values may not be achieved.

At the end of the meetings, the jurors are likely to have learnt a great deal about the relevant issues and be expected to reach a conclusion on the matter considered, preferably based on consensus but, if not, through the use of a voting procedure. Unlike a real jury, in most cases the decisions made within citizens' juries will be non-binding. As with stated preference, where there may be some uncertainty regarding the link between response and policy, this problem is also present in citizens' juries (Ward, 1999). Some degree of consequentiality can be ensured using a 'contract' signed between the commissioning authority and participants that the outcome of the jury will be taken seriously within future decision making. Implications of failing to achieve a level of consequentiality were demonstrated within the study by Aldred and Jacobs (2000). Upon realizing the lack of clarity between the jury verdict and policy, Aldred and Jacobs (2000) report expressions of lack of trust and suspicion being stated. Indeed, strategic behaviour was observed within the making of the verdict.[7] Again these results suggest the difficulties of using the jury metaphor.

As with the use of other group methods, the tape recording of the citizens' jury process provides a great deal of information on the process by which people make their decisions and the views and level of understanding of the participants. This is likely to provide a wealth of information beyond the actual decision made. In the case that the verdict is used as a justification for the final policy outcome, this information will be particularly useful as a means of persuading others that the citizens' jury process is fair, unbiased and sufficiently detailed such that the judgements were made by 'informed' citizens. In the case where the final policy decision differs from that of the jury, understanding the reasons for the verdict will be particularly useful in helping devise ways for workable agreements to be made such that conflicts can be addressed within actual policy development. The qualitative information can also be important if the jurors failed to reach a consensus, as the reasons for this failure can be considered and appropriate changes made to the policy such that a consensus would be more likely.

INDIVIDUAL AND SOCIAL WILLINGNESS TO PAY

Given the communal nature of environmental scenarios and payment vehicles considered, it would seem important that stated preference does not occur in social isolation. In the last chapter a citizen's commission approach was described, which enabled social interaction to occur prior to making individual valuations. The idea behind this metaphor is that 'a representative sample of citizens is selected to learn about an issue on behalf of the electorate' (Fischhoff, 1997: 199). This is also a key principle within citizens' juries, within which citizens become 'informed' through the calling and cross-examination of witnesses. The use of witnesses differs from the attempts to provide neutral information within stated preference, as the witnesses may take particular positions regarding policy choices. A further difference is the form of valuations elicited. A number of researchers have suggested that instead of the aggregation of preferences elicited individually, a social willingness to pay should result from consensus following a process of free and open debate (Ward, 1999; Gregory and Wellman, 2001; Wilson and Howarth, 2002; Howarth and Wilson, 2006). This suggests there are two types of values: (a) whether the individual is willing to pay the amount for the goods and services provided; and (b) whether society as a whole is willing to pay. As noted above, Howarth and Wilson (2006) suggest equality between these values is only achieved in the unlikely occurrence that the individual preferences for the public good are perfectly aligned with the existing tax system.

An objection to stated preference is concerning the implicit value judgement that those with the most money have the greatest say in the outcomes of cost-benefit analysis or referendum-type decision rules. It has been noted above that this approach is not considered to lead to a socially equitable outcome (Wilson and Howarth, 2002). Using a consent-based approach for estimating SWTP, every member, in theory at least, has an equal say, where it is the quality of their arguments rather than the status (financial or otherwise) of the participants that counts.[8] Although there is a loss of statistical representativeness using this approach, a social willingness to pay based on mutual consent from the group would appear to be fairer in the sense of equal say[9] and more consistent with a democratic system.

An argument in favour of using the individual stated preference approach (conventional or following a group meeting) relates to the making of a financial commitment. It could be argued that as preferences are elicited on potential financial payments, these are best considered in an individual setting. This problem can be alleviated within a group session where questionnaires are self-completed following the discussion and respondents are not obliged to divulge issues of a 'private' nature. This would represent an improvement on the conventional stated preference, as the lack of an interviewer is less likely to encourage bias due to the social context of the interview. In addition, the use of confidential votes could be argued to be more consistent with democratic ideals. Within a citizens' jury, in theory at least, such personal information would have little relevance as the jurors are meant to consider issues in terms of society as a whole. Indeed, within a group-based process, it is more likely that in order to convince others to your way of thinking, arguments will need to be expressed in societal rather than individual terms (Goodin, 1986; 1992; Elster, 1998). Although the jurors can consider the effects of policy on minority groups and social equity, individual desires of the participants would need to be eliminated from the process. In a real jury this is possible as the participants are unaffected by the outcome, but as mentioned previously this is not always the case within citizens' juries. The focus of citizens' juries, however, is such that social pressure ensures, to a certain degree, that societal issues remain the focus of the debate. This is particularly the case when considering non-use values (Jacobs, 1997).

A key advantage of the social WTP approach is that consensus is more likely to be achieved. Although not stated in social isolation, individual confidential votes elicited at the end of the meetings are less likely to lead to clear outcomes. As such, social WTP may be more likely to provide an outcome which has the consent of the group and the qualitative findings from the discussion will illustrate how this consensus was reached. This may illustrate how a real outcome can be achieved through stakeholder negotiation. However, the link between using a consensus approach and the likelihood of agreement has not been formally tested. It is also important to test whether the longer approach of a citizens' jury, compared to that of short group meetings without witnesses, is more likely to lead to a consensus. The citizens' jury approach is more time consuming and expensive and perhaps less representative (can undertake fewer meetings) than the group-based citizens' commission suggested in Chapter 7. Whether the citizens' jury process is more likely to lead to a consensus and whether decisions are improved is unclear.

Although WTP can be used in a referendum format, it is common to estimate values as an input into cost-benefit analysis. Within Chapter 5 concerns were raised that some respondents may have used heuristics inconsistent with maximum WTP and that respondents were commonly considering costs (though actual cost was not given) within their responses. These findings complicate the use of WTP within cost-benefit analysis. In the case of social WTP participants are being encouraged to make judgements based on the wider interests of society as a whole. Within such decisions it is only natural that participants consider aggregate societal costs and benefits, making it inappropriate to use social WTP within cost-benefit analysis. James and Blamey (2005) suggest instead of being incorporated within cost-benefit analysis social WTP could be used in terms of social welfare, where the results could be:

> interpreted as the maximum amount of money that could be charged to members of the public using a given payment vehicle in order to obtain the posited environmental programme without leaving social welfare as a whole worse off than prior to the change. (James and Blamey, 2005: 239)

The social WTP could be assessed through the aggregation of a per person average to society as a whole or as in the case of James and Blamey (2005) through the revenue generated through a form of income tax. Although the form of social WTP may vary between studies, the estimation of social welfare would provide a useful comparison to the cost of the schemes. It is important to note that, within the design of the elicitation methods it is necessary to describe how the scheme would be funded and the implications in terms of who pays and how much. Although some degree of choice as to the most appropriate means of payment can be left to the jury, this illustrates the difficulties associated within payment vehicles and that attitudes towards the authority responsible for delivery remain.

PRACTICAL APPLICATIONS

Although much discussed within the literature, there have been very few actual studies estimating social WTP. Indeed, the author is only aware of four studies that have elicited WTP using a citizens' jury type method. These will be discussed briefly in turn, before considering the implications of the findings.

Valuation Within a Citizens' Jury: National Park Management

James and Blamey (2005) used a citizens' jury to estimate WTP to help decide between management options for national parks in New South Wales, Australia. Prior to undertaking the citizens' jury, national park management activities were reviewed within a series of focus groups and from this an explanatory document was developed to provide to the jurors. A status quo situation was identified and described using attributes in the choice experiment format in terms of fire management, feral animal control, weed control, maintenance of facilities and protection of historic sites. Three other options were identified, two of which were within the current budget and one which would cost more. The jurors were told that this extra expenditure could be raised using a park management levy to be paid by all residents of the region of New South Wales. The jurors were told that the cost of this fourth option was unknown, but would cost substantially more than the existing budget. In addition to prewritten information regarding the park management and the role of jurors within the jury, information was provided to the jurors through seven witnesses representing expertise in the management attributes considered. The jury lasted three days and the process was helped by the use of a facilitator. There was no direct link between the verdict of the jury and policy development, however, the jurors were told that the results would be given to the park service for use within budgeting and in design of public participation programmes.

The jurors were asked to complete two tasks. The first was to consider the status quo situation to that of the alternative management options within the existing budget. This was done privately initially where jurors were asked to provide the reasons for their choice. The results were then displayed anomalously and provided the starting point for the debate. Although agreement was not initially reached, consensus was eventually achieved through noting the concerns of those initially choosing other options. The second task was to consider the improved management package, where the jury were asked to assess: 'how high would a park management levy have to be before the jury would recommend Option 1 [status quo option] rather then Option 4 [the improved option]?' (p. 231).

After much deliberation, no consensus was achievable with three of the 13 jurors maintaining their objection to the levy. On the basis of the comments of the other jurors social WTP was estimated. The jurors provided recommendations in terms of how the levy should be collected. As is common within conventional stated preference studies (Chapter 5), the jurors encountered difficulties with the notion of maximum WTP and the cost of the scheme. James and Blamey (2005) report this challenge to relate to the question: 'if we don't know how much it is going to cost for each of those (activities), how do we know the amount of the levy to impose to cover the costs?' (p. 235). Other challenges encountered related to the degree of juror participation. James and Blamey (2005) report, that of the 13 jurors, only one spoke when directly questioned and several others made minimal contributions to the discussions. This was seen to be partly due to the wide range of 'social and cognitive skills' of the participants, which is required in order for the jury to reflect the population considered (p. 237).

Using Trade-offs to Help Develop an Estuary Improvement Plan

In the last chapter the multi-attribute utility approach developed by Gregory *et al.* (1993) was introduced, which takes a deliberative approach to preference construction within which the holistic scenario is subdivided into a list of additive components which people desire. Gregory and Wellman (2001) provide an application of this approach which estimated social WTP for various actions which could be implemented for an estuary project in Oregon, USA. Such action was to form part of a conservation and management plan for the area. Although other methods of participation had been tried (public meetings), Gregory and Wellman (2001) suggest that:

> although these approaches can provide useful general input, the pressing need for the TBNEP [estuary improvement project] was to find a way to involve local residents meaningfully at a detailed, action-specific level and to ensure that the judgements of participants were informed by and recognize the complex, multidimensional nature of the types of program initiatives under consideration. (p. 41)

Through consultation with project staff, community leaders and stakeholders, a means-ends network was developed and agreed, which clearly depicted the actions that could be taken within the project and which objectives they would help achieve. The second stage of the study was to gain an understanding of the trade-offs involved with the implementation of alternative actions. Following further consultation, three ecosystem management options were identified for further consideration and a workbook developed which considered the trade-offs using an approach in a matrix similar to that used by choice experiments. Within the workbooks, group participants were asked first to consider the three actions identified and then to consider only one of these actions in detail. Social WTP was elicited and justified in terms of concerns regarding charitable donations and perceived closeness with actual economic opportunity costs and public awareness. Within the choice cards, an attribute was added representing the aggregated costs of the schemes and social WTP was determined by upper (maximum agreed to) and lower (minimum agreed to) costs. The results were seen to be cognitively valid, as the workbooks were completed within small groups, co-led by facilitators. These groups allowed focused discussions and questions to be asked concerning any missing facts.

Comparing Participatory and CV Responses

For a forest floodplain restoration case study in the Scottish borders, Kenyon and Hanley (2005) compared the results of a citizens' jury with those of a conventional CV study and also a 'valuation workshop' that attempts to combine the advantages of both approaches.[10] In the context of forest floodplain restoration, the citizens' jury considered the objectives of land use and environmental projects in southern Scotland and the ways the success of such projects should be determined.

The jury consisted of 11 jurors that were recruited from those undertaking the CV study. The meeting lasted three days and involved ten witnesses. Techniques suggested by Petty *et al.* (1995) were used to help engage participants. The jury were in favour of the specific project and suggested that it and other similar projects should be judged in terms of a series of environmental and social criteria. Interestingly, economic criteria were not raised and no valuation was undertaken within the citizens' jury. The CV survey used a charitable donations payment vehicle and, from the sample of 336 respondents found an average WTP of £13. Interpreting these results in terms of the forest project considered, it was found to pass the cost-benefit test (benefits £570 000 and costs £350 000). A relatively high number of responses (29 per cent) were protest bids, suggesting some concern regarding the accuracy of the WTP estimates. There was a degree of uncertainty regarding respondent attitudes towards the scheme, with 13 per cent unsure if they preferred the area with or without the scheme. This problem suggested that the information provided within the survey might not have been sufficient. Interestingly, the most significant variable affecting WTP was whether the respondent was likely to visit the area, suggesting a more consumer rather than citizen preference.

The valuation workshop was undertaken separately, using a mail recruitment method. The exercise consisted of 44 participants involved in four workshops, each lasting three hours. As the workshop used the same presentation as the CV survey, the participants were not given control over their information requirements. However, participants were able to discuss the issues in a social setting and, as with the citizens' jury, 'best practice' methods were used to encourage participation. The CV questionnaire was filled in at the beginning of the meeting and participants revisited their responses again at the end of the meetings. The results of the CV questions differed marginally from those of the more conventional CV survey (median significantly different but not mean), with a lower percentage of protest bids but a higher percentage of 'don't knows'. The effect of the meetings was to cause 14 per cent to change their mind, all positively affecting the final WTP estimate. The information problem was marginally reduced as two of the participants changed their 'don't knows' to positive bids. Reaction to the valuation exercise was not very positive, with some participants suggesting that it is not possible to put a value on such a project, whereas others raised concerns over the realism of the estimates, objecting to the payment vehicle used[11] and raised feelings that information regarding cost was required in order to consider the value of the scheme. These results are not untypical to those reported in Chapter 5. The valuation workshops also provided a range of qualitative information of a similar type to that of the citizens' juries. This demonstrates how the workshops provide a combination of the two approaches as well as a better understanding of the meaning of the valuation estimates.

Comparing Citizen and Individual Choice Experiment Responses

The final case study is provided by Álvarez-Farizo and Hanley (2006), who explored how the results of choice experiments changed as participants are involved in a 'valuation workshop' in which they are given more time to think, an opportunity to discuss the issues with others (including family and friends), and move from an individual to a collective approach. The comparison of valuations took place over a three day event. On the first day of the event participants undertook the choice experiment in survey conditions in a self-interest perspective followed by a discussion. On the second day participants were encouraged to debate the issues further as if they were choosing on behalf of the community and again completed the same questionnaire. Finally, on the third day, as well as discussing the issues further, the choice experiment was undertaken collectively rather than individually, choosing the options with the highest votes. Comparing the parameter estimates in the choice experiment models, a significant difference was observed. Likewise, a number of the implicit price estimates were statistically different under individual and citizen perspectives. The results suggest that preferences change when people are given more information, and time and that moving from an individual to a collective/citizen perspective changes both values and preferences. Although choice of values is likely to affect the policy recommendations, in the absence of a means of judging between different valuations, Álvarez-Farizo and Hanley (2006) left the reader to question as to which approach generates 'better' answers.

CONCLUSIONS

This chapter has considered whether lessons learnt from citizens' juries and the use of group methods can improve the social aspects of the valuation process. These issues have been demonstrated through four case studies that illustrate the range of approaches and problems encountered when ensuring the inclusion of citizen type values within decision making. Compared to stated preference surveys, consent-based approaches perform better in terms of the depth of analysis provided, enabling plural values to be elicited, the selection of equitable outcomes, as well as a more active engagement with the issues being debated. However, the benefits have been at the expense of a range of the population involved in the process, where a much smaller number of individuals are involved. This is particularly the case in terms of citizens' juries, where there is little hope of achieving a statistically representative sample and the length of the juries is such that even if financial compensation is provided very few full-time employed people are likely to be able to take part. Although it can be argued that the aim should be a representation of discourses rather than individuals, the problem does not go away. There are also serious concerns in terms of involvement amongst participants. Bringing together a socially and economically mixed group is not consistent with the best practice within focus groups and is likely to be daunting for some members. Although not assessed within the other studies, James and Blamey (2005) illustrate the difficulties of gaining participation from a 'representative' group of varied backgrounds.

More generally, a number of the concerns of stated preference are not alleviated by the use of consent and group-based methods. Issues relating to the payment vehicle used and the organization responsible for delivery remain, where trust can still play a major part in the analysis. Strategic behaviour was also observed within the study by Aldred and Jacobs (2000). As the verdict of citizens' juries is unlikely to be binding, this demonstrates the virtual impossibility of devising a method that is not hypothetical. Effectively, decisions will be made through representative democracy. Given the complexity and uncertainties concerning environmental issues as well as the imperfect nature of any method suggested, it may be considered inappropriate for the results from public consultation to directly determine the policy outcomes.

The results of this and the previous chapter have illustrated the importance of emphasizing understanding as well as valuation. Case studies have demonstrated the breadth of understand that can be achieved through the use of group methods and despite not being a panacea, such methods have demonstrated their worth in terms of giving meaning to valuations elicited. The social/private dilemma will remain. Whether using individual interviews or consent-based group methods, the values elicited will almost always involve issues relevant to the individual and society as a whole.

NOTES

1. Through achieving a consensus within a 'representative' group, a consent-based approach is seen to gain the consent of the general public to undertake the scheme considered at a given cost borne by the tax payer.
2. Such flexibility does not enable, however, a consistent policy when making environment decisions.
3. This was also a cause of the protests reported by Stevens *et al.* (1991).
4. This section focuses on citizens' juries because of their potential relevance to stated preference research, but it should be noted that there are other deliberative democracy approaches which may be of use within public participation on environmental issues. See Smith (2003) for details.
5. Even with payment, getting people to use their holidays to attend such a meeting is likely to be difficult and likely to bias the characteristics of those attending the meetings.
6. Given the costs of undertaking citizens' juries this is likely to prohibit the undertaking of multiple groups.
7. The importance of maintaining trust within citizens' juries has been widely stated (Coote and Lenaghan, 1997; Ward, 1999; Aldred, 2005).
8. Those with more status may be better able to articulate their views.
9. It could also be argued that those paying the most tax should have the greatest say.
10. See also Kenyon *et al.* (2001) for a more detailed discussion of the citizens' jury and Kenyon and Nevin (2001) for a discussion of both the CV study and the citizens' jury.
11. The charitable donation payment vehicle was decided as the most applicable with pre-survey focus groups. As such, this issue was always going to be controversial.

Conclusion

9. Evaluating and redesigning stated preference valuation

INTRODUCTION

Environmental issues are inherently intractable and complex. Knowledge is contested amongst experts with debates in terms of ethics and principles as well as contests between interests in the pursuit of sustainability. When involving the general public in environmental consultation they are faced with a difficult cognitive task, having to deal with uncertainty of outcomes, lack of clarity as to the best environmental outcome, difficulties of demarcation and a plurality of values which may be incommensurable. The task of the researcher is to reduce this cognitive load, but also to maintain sufficient understanding such that meaningful responses can be elicited. One form of public consultation that has attempted to address these difficulties is questionnaire-based stated preference methods, the valuations from which may be used to feed into cost-benefit analysis, enabling a formal consideration of the efficiency of resource use. These approaches have been developed in response to numerous market and intervention failures, within which insufficient consideration has been given to the non-market benefits of wetlands, forests and other natural assets.

Despite the potential of stated preference methods, some commentators have seriously questioned the effectiveness of stated preference methods. Given the extent of criticism, no exhaustive categorization of challenges facing stated preference practitioners can be made. However, a framework for interpreting stated preference responses has been developed based on criticisms from economics, behavioural psychology and more social/institutional approaches. There are three recurring themes within the literature that have been noted: the cognitive task faced by respondents; the hypothetical nature of the transaction; and the communal nature of the scenarios considered.

The key implication of the cognitive task is that respondents are seen to construct their preferences within stated preference surveys. Behavioural psychology literature has demonstrated how such preferences are unstable, sensitive to task complexity, time pressure, response mode, framing, reference points and numerous other contextual factors. Though some of these claims have been contested or seen to be consistent with economic theory, the evidence suggests they are difficult to eradicate. To economists the hypothetical nature of the choices made within stated preference are a key concern and as a general principle, it is crucial that the perceived linkages between response and policy-formation are sufficient for respondents to take the price seriously and put in the required effort to give meaningful answers. Recent evidence has suggested that such hypothetical biases can be reduced through reminders to consider the personal financial consequences of their responses. However, combining the task complexity and the hypothetical nature of the scenarios, there is a need not only for respondents to take the amounts seriously, but also that the price is sufficient for respondents to put in the required effort. Lastly, the communal nature of environmental decisions suggests they are social as well as individual in nature and that there may be moral and ethical issues involved. These issues complicate the meaning of stated preference responses and may lead to a high refusal rate. Indeed, the communal nature of the scenarios may mean that respondents are torn between responding as citizens or as consumers, where there may be incommensurability between the two. This book has explored the extent to which these challenges facing stated preference methods can be overcome through the use of mixed methodologies. Redesigning stated preference using mixed methods two approaches are considered, first in terms of mixing methods within conventional stated preference; and second through moving away from the conventional approach to explore the use of group methods within preference construction and forming a social consensus on willingness to pay.

MIXING METHODS WITHIN CONVENTIONAL STATED PREFERENCE

Using conventional stated preference methods, environmental valuations are elicited within an individual questionnaire format using either market or referendum-type methods. By mixing methods initial exploratory qualitative analysis can be used to explore shared understanding in terms of perceptions, categories and language, as well as how these differ from those used by experts. Furthermore, structured qualitative analysis can then be used to test the questionnaire (understanding and interpretation of the questions, sensitivity to deliberation) and explore the meaning of responses. Further extending this process, a list of attitudinal questions can be developed from the qualitative findings which enable quantitative testing of the extent to which the issues raised within the qualitative results are also representative of the wider population. The public acceptability of the stated preference approach adopted can also be considered. Qualitative analysis can, where necessary, also be used post-survey to explore empirical regularities or anomalies encountered within the quantitative results and any adjustments required for issues of, for example, equity.

Within Chapters 5 and 6 the findings of qualitative analysis and their implications for stated preference methods have been given detailed consideration, where the issues of cognitive limitations, hypothetical payment and the communal nature of the scenarios and payment vehicles are all seen to have relevance.

- *Presentation:* difficulties encountered within the presentation of the scenarios (including payment) have demonstrated the inevitable limitations of the personal interview in terms of the time needed to digest the information and the complexities of the scenarios considered. Although a level of information can be provided that satisfies most of the respondents, they may also misconstrue this information, either through their misunderstanding or mistrust of the information presented. Such problems question whether respondents can adequately construe the pertinent information and whether there can be sufficiently consistent within respondent understanding such that the valuations are meaningful.
- *Payment vehicles:* the results of the qualitative analyses demonstrate that funding adds an additional dimension to the meaning of the responses and complicates analysis. It was demonstrated that an 'ideal' payment vehicle may not always exist, where even the choice of the payment vehicle that is most popular with the respondents may still lead to problems of respondents denying responsibility for payment; a lack of trust in the authority responsible for provision; and confusion as to the context of the payment. Although such findings may be consistent with economic theory, it is difficult to gain an understanding of these effects and the interpretation of valuations is complex.
- *Strategies and motivations*: The findings from qualitative studies suggest that ability to pay, price, quantity and quality only provide a partial understanding of the strategies and motivations used when responding to stated preference questions, with social, cultural and ethical factors also playing their part. As such the resulting behaviour is much more difficult to predict and generalize than neoclassical economics assumes.
- *Acceptability of the approach*: the qualitative results have suggested that for some scenarios it is possible to gain a majority in favour of the use of stated preference, even when informed in detail of the method being used and its proposed use within policy making. These results have indicated that it is not sufficient for decision makers to simply ask for the value the general public place on a scenario, rather, they need to also ask what kind of values they are able/willing to articulate and whether the method is appropriate for the scenario considered.

Overall these results illustrate the complexity of valuing environmental goods. The qualitative results are mixed, and interpretation of the implications for stated preference methods will depend on the researchers' attitude towards the ability of respondents to articulate their preferences, and the need to make decisions in a social setting and/or with real economic payments. Whatever the researchers leaning regarding these issues, the results also suggest that the problems identified can be reduced through good design, selective choice of scenarios that are not too complex, and with which respondents have a high level of engagement, and/or have consistency between the payment vehicle and the scale of the goods and services valued. The degree to which these issues can be met will affect the quality of the valuations elicited.

Consistent with the argument generated throughout this book, if stated preference surveys are going to be undertaken, the use of qualitative methods are essential and need to be integrated within valuation surveys in order to improve design and help the interpretation of responses. Due to the presence of complexities and the potential for misinterpretation, a failure to employ sufficient resources may lead to the incorrect application of values within policy making. As such, prior to undertaking a stated preference survey it is recommended that a feasibility study is undertaken into the extent to which the scenario is adequately developed and the applicability of stated preference methods to the task.

Developing an Environmental Project

Using stated preference surveys as a means of exploring how to develop environmental schemes and their funding is very expensive. Instead a more qualitative approach would be more cost effective and probably more rewarding. Furthermore, if the scenario and its funding are not properly developed prior to undertaking a stated preference study, the valuations may have little meaning in terms of the final scheme developed. Although using choice experiments at least introduces some flexibility, the application of values to policy may still be limited. The results of pre-survey qualitative methods have illustrated the range of information necessary to develop a scheme which is likely to be most acceptable to the general public, in terms of the payment vehicle, developing trust in the agency responsible for delivery as well as issues for which respondents may have ethical difficulties. The use of trade-offs within focus groups, for example, can be used to explore these issues effectively.

Applicability of Stated Preference

Further to policy design, it is also crucial that the applicability of the stated preference be assessed prior to undertaking a major survey. The study by Vadnjal and O'Connor (1994) provides a classic example of a valuation study where the prior use of qualitative methods would have identified that the scenario was not applicable to the use of stated preference. It is important to start with a feasibility study and if the approach is found to be 'fit for purpose' then proceed. The study by Vadnjal and O'Connor (1994) does however, demonstrate that the use of trade-offs within stated preference can provide a very useful tool within qualitative methods for understanding how the general public relate to an environmental issue. In order to make a judgement regarding applicability it is perhaps necessary to have a set of criteria. The issues outlined in Chapter 5 around presentation, payment vehicles, strategies, motivations and acceptability of the approach to the general public would represent a good starting point for the development of judgement criteria. If a given scenario is not considered applicable for stated preference methods, it may be necessary to move away from conventional approaches. However, it is important to remember that alternative approaches may reduce the validity of valuations for use within cost/benefit analysis and the results may be judged to be less representative of the population of interest. Given the complexity of environmental scenarios and their communal payment, a degree of divergence between the idea and the applicability of the actual scenario will need to be accepted with any approach to consultation considered. The use of detailed attitudinal statements enables some control and understanding of the issues of concern.

MOVING AWAY FROM THE CONVENTION

Chapters 7 and 8 have considered the use of group methods within the construction of preferences and, through the discussion with others, consideration of social, cultural and ethical factors. These were explored through the perspective of the citizen as well as the individual. By way of a summary, Table 9.1 provides suggestions of how the three issues of cognitive ability of respondents, hypothetical nature of the transaction and communal nature of the scenarios can be alleviated. Although the use of qualitative methods provides little help with concerns regarding the hypothetical nature of the transactions, recent results from real payment testing suggests a potential way forward (See Chapter 5).

Cognitive Ability of the Respondents

In the presence of evidence suggesting an inadequacy of the conventional interview situation, potentially a group-based approach could be used, with a follow-up meeting to allow participants the opportunity to discuss the issues with family and friends. Such groups provide a permissive, non-threatening environment that is conducive to a relaxed discussion. They also enable participants themselves to decide what they need to know and ask for any additional information they require.

Table 9.1 Assessing the applicability of stated preference methods

	Cognitive ability of the respondents	Hypothetical nature of the transaction	Communal nature of the scenarios considered
How to improve preference elicitation?	Only ask questions that respondents can answer well or design an approach in which respondents can better construct their preferences	Design an instrument that has a clearer response-policy link and better mimics a market or referendum situation	Adopt a more social approach to consultation which allows the consideration of shared values and equity

Considering the evidence presented in Chapter 7, in the case of unfamiliar scenarios the use of the group method produces different, and in most cases 'better' quality valuations, to those conducted through individual interviews. Particularly evident is that respondent understanding of the issues is superior to that achieved within a conventional individual interview. Providing follow-up meetings or interviews after a time interval would also appear to improve respondent understanding and the valuations elicited.

As the use of group methods breaks conventional survey norms, its adoption does not come without methodological implications. Specifically, the use of the group approach no longer maintains cross-respondent consistency in the information provided; may lead to a different focus on communal issues; and there are questions regarding the ability of using such an approach to conduct a representative sample of the population of interest. Although using the group approach does appear to improve valuations, it only alleviates the problems and in the case of highly complex and unfamiliar goods and services these improvements may still be insufficient to generate stable responses. However, the group approaches, although imperfect, do lead to an enhanced understanding of general public attitudes and preferences for the goods and services provided. These data are likely to be informative within the decision making process for the schemes considered.

Where the complexities of the situation considered make it inevitable that unstable values will be elicited, the possibility of using alternative approaches which adjust existing valuations made in more favourable circumstances have been considered. The potential for achieving this was explored in Chapter 7 in terms of using group based methods with experts or the general public. Examples are provided that suggest that such an approach is by no means simple and should only be undertaken with care.

Communal Nature of the Scenarios Considered

Chapter 8 considered whether lessons learnt from the use of consent-based group approaches such as citizens' juries can be adapted to help deal with the communal nature of the environmental scenarios and payment vehicles used. Compared to stated preference surveys, consent-based approaches perform better in terms of the depth of analysis provided, enabling plural values to be elicited, the selection of equitable outcomes, as well as a more active engagement with the issues. However, there is a clear loss of representation of the general public. This is particularly the case in terms of citizens' juries, where there is little hope of achieving a statistically representative sample and the length of the juries is such that even if financial compensation is provided only those with time are likely to be able to take part. Although it can be argued that the aim should be a representation of discourses rather than individuals, the problem does not go away. Bringing together a socially and economically mixed group is also not consistent with the best practice within focus groups and is likely to be daunting for some members. This will further reduce the representativeness of the meetings.

More generally, consent and group-based methods are susceptible to similar challenges as stated preference. The issues relating the payment vehicle identified in Chapter 5 remain. Furthermore, citizens' juries are unlikely to be binding, such that they are hypothetical in nature. The social/private dilemma also is still also present. A judgement needs to be made within the feasibility survey as to the comparative importance of these issues related to the specific study considered. If the purpose of the research is to provide benefit estimates for cost-benefit analysis, the consent-based approaches are unlikely to be appropriate.

CONSIDERING THE FUTURE FOR STATED PREFERENCE

A key motivation for this book was the authors' previous confusion over the way forward for environmental valuation. It was decided that in order to develop a valid opinion it was necessary to discover what the key difficulties were, consider the alternative approaches and consider which were the most appropriate and in what circumstances. This process was very satisfactory, in that the book has achieved these objectives, however, the answer as to the way forward, as the discipline, has turned out to be more complex than expected.

The analysis has demonstrated how, given the complexities of the scenarios and payment, pluralities of value and the inevitable hypothetical nature of the questions, there is not one single method, or ever likely to be, that simultaneously solves these problems. This concern has been recognized within the literature, where, for example, Jacobs (1997) suggests there is not an ethically neutral approach to consultation on these issues and Niemeyer and Spash (2001) suggest:

> No one particular approach can address all problems, nor is any one particular approach necessarily wrong in all circumstances. Ultimately, a variety of approaches are needed but this means recognising when and how particular tools are unsuitable. (p. 583).

Inevitably a flexible approach is required for method selection, which is consistent with the suggestion above of an initial feasibility study to be undertaken prior to deciding how to proceed and the type of information that can be gained from consultation. Guidelines are required as to under which circumstances different approaches are to be applied. This book has provided a starting point from which such guidelines can be developed. The challenges outlined also suggest that it is not possible to design a standard approach, such as extended cost-benefit analysis, as an input into all major environmental decisions. Likewise, the guidance provided by the NOAA panel (Arrow *et al.*, 2003), would need to be much broader to incorporate the range of methods which can be used. The use of stated preference reported in the literature varies considerably in terms of the way the information is presented and the form elicitation methods take. This is even the case with the referendum style dichotomous choice recommended by the NOAA panel. There is a danger that the need to follow a study specific approach will lead to a failure to follow standards and ensure robustness in the techniques used.

As environmental issues are inherently intractable and complex, the valuation of environmental goods and services will be similar in character. Despite the criticism within the literature, undertaking mixed approaches to stated preference much enhances our understanding of general public preferences regarding environmental issues. Although the conventional approach will increasingly become only one of a number of means of assessing environmental values (social and private), it still has a future and remains a worthy area for future research.

References

Adamowicz, W.L., J.J. Louviere and M. Williams (1994), 'Combining revealed and stated preference methods for valuing environmental amenities', *Journal of Environmental Economics and Management,* 26, 271–92.

Adamowicz, W., P. Boxall, M. Williams and J. Louviere (1998), 'Stated preference approaches for measuring passive use values: choice experiments and contingent valuation', *American Journal of Agricultural Economics*, 80, 64–75.

Ajzen, I. (1991), 'The theory of planned behaviour', *Organizational Behavior and Human Decision Processes*, 50, 179–211.

Ajzen, I. and M. Fishbein (1977), 'Attitude behaviours relation: a theoretical analysis and review of empirical research', *Psychological Bulletin*, 84, 888–918.

Ajzen, I. and B.L. Driver (1992), 'Contingent value measurement: On the nature and meaning of willingness to pay', *Journal of Consumer Psychology*, 1, 297–316.

Ajzen, I., T.C. Brown and L.H. Rosenthal (1996), 'Information bias in contingent valuation: effects of personal relevance, quality of information and motivational orientation', *Journal of Environmental Economics and Management*, 30, 43–57.

Ajzen, I., T.C. Brown and F. Carvajal (2004), 'Explaining the discrepancy between intentions and actions: the case of hypothetical bias in contingent valuation', *Personality and Social Psychology Bulletin*, **30**(9), 1108–21.

Alberini, A., B.J. Kanninen and R.T. Carson (1997), 'Modeling response incentive effects in dichotomous choice contingent valuation data', *Land Economics*, **73**(3), 309–24.

Aldred, J. and Jacobs, M. (2000), 'Citizens and wetlands: evaluating the Ely citizen's jury', *Ecological Economics*, 34, 217–32.

Aldred, J. (2005), 'Consumer valuation and citizen deliberation', in M. Getzner, C.L. Spash and S. Stagl (eds), *Alternatives for Environmental Valuation, Routledge Explorations in Environmental Economics*, London: Routledge.

Álvarez-Farizo, B. and N. Hanley (2006), 'Improving the process of valuing non-market benefits: combining citizens; juries with choice modelling', *Land Economics*, **82**(3), 465–79.

Andreoni, J. (1990), 'Impure altruism and donations to public goods', *Economic Journal*, 100, 464–77.

Arrow, K.J. (1982), 'Risk perception in psychology and economics', *Economic Enquiry*, 20, 1–9.

Arrow, K.J. (1986), Comments within Chapter 12 of R.G. Cummings, D.S. Brookshire and W.D. Schulze (eds), *Valuing Environmental Goods: An Assessment of the Contingent Valuation Methods*, Totowa, NJ: Rowman and Allanheld.

Arrow, K.J. and A.C. Fisher (1974), 'Environmental preservation, uncertainty and irreversibility', *Quarterly Journal of Economics*, 88, 312–19.

Arrow, K.J., R. Solow, P.R. Portney, E.E. Leamer, R. Radner and H. Schuman (1993), *Report of the NOAA Panel on Contingent Valuation*, Washington, DC: National Oceanic and Atmospheric Administration, 11 January.

Baker, R. and A. Robinson (2004), 'Responses to standard gambles: are preferences "well-constructed"?' *Health Economics*, **13**(1), 37–48.

Baron, J. and J. Greene (1996), 'Determinants of insensitivity to quantity in valuation of public goods: contribution, warm glow and budget constraints, availability and prominence', *Journal of Experimental Psychology: Applied*, **2**(2), 107–25.

Barr, S., N.J. Ford and A.W. Gilb (2003), 'Attitudes towards recycling household waste in Exeter, Devon: quantitative and qualitative approaches', *Local Environment*, **8**(4), 407–21.

Bateman, I.J., M. Cole, P. Cooper, S. Georgiou, D. Hadley and G.L. Poe (2004), 'On visible choice sets and scope sensitivity', *Journal of Environmental Economics and Management*, **47**(1), 71–94.

Bateman, I.J., I.H. Langford, A.P. Jones and G.N. Kerr (2001), 'Bound and path effects in double and triple bounded dichotomous choice contingent valuation', *Resource and Energy Economics*, 23, 191–213.

Bateman, I.J., I.H. Langford, R.K. Turner, K.G. Willis and G.D. Garrod (1995), 'Elicitation and truncation effects in contingent valuation studies', *Ecological Economics*, 12, 161–79.

Beggs, S., S. Cardell and J. Hausman (1981), 'Assessing the potential demand for electric cars', *Journal of Econometrics*, 17, 1–20.

Ben-Akiva, M., T. Morikawa and F. Shiroishi (1991), 'Analysis of the reliability of preference ranking data', *Journal of Business Research*, 23, 253–68.

Bennett, J. and R. Blamey (eds) (2001), *The Choice Modelling Approach to Environmental Valuation*, Cheltenham, UK and Northampton, MA, US: Edward Elgar.

Berg, B.L. (1995), *Qualitative Research Methods for Social Scientists*, 2nd edn, London: Allyn and Baron.

Bergstrom, J.C., J.R. Stoll and A. Randall (1989), 'Information effects in contingent markets', *American Journal of Agricultural Economics*, 71, 685–91.

Blamey, R.K. (1996), 'Citizens, consumers and contingent valuation: clarification and the expression of citizen values and issue-opinions', Chapter 7 in W.L. Adamowicz, P.C. Boxall, M.K. Luckert, W.E. Phillips and W.A. White (eds), *Forestry, Economics and the Environment*, Oxford: CAB International.

Blamey, R. (1998), 'Contingent valuation and the activation of environmental norms', *Ecological Economics*, 24, 47–72.

Blamey, R., M. Common and J. Quiggin (1995), 'Respondents to contingent valuation surveys: consumers or citizens', *Australian Journal of Agricultural Economics*, **39**(3), 263–88.

Blamey, R., J. Gordon and R. Chapman (1999), 'Assessing environmental values of water supply options', *The Australian Journal of Agricultural and Resource Economics*, **43**(3), 337–57.

Blamey, R.K., R.F. James, R. Smith and S. Niemeyer (2000), 'Citizens' juries and environmental value assessment', Australian National University, Canberra, accessed at http://cjp.anu.edu.au/docs/CJ1.pdf

Bogardus, E.S. (1926), 'The group interview', *Journal of Applied Sociology*, 10, 372–82.

Bohara, A.K., M. McKee, R.P. Berrens, H. Jenkins-Smith, C.L. Silva and D.S. Brookshire (1998), 'Effects of total cost and group-size information on willingness to pay responses: open-ended vs. dichotomous choice', *Journal of Environmental Economics and Management*, 35, 142–63.

Boyle, K.J., W.H. Desvousges, F.R. Johnson, R.W. Dunford and S.P. Hudson (1994), 'An investigation of part-whole biases in contingent valuation studies', *Journal of Environmental Economics and Management*, 27, 64–83.

Boyle, K.J., F.R. Johnson, D.W. McCollum, W.H. Desvouges, R.W. Dunford and S.P. Hudson (1996), 'Valuing public goods: discrete versus continuous contingent valuation', *Land Economics*, **72**(3), 381–96.

Boyle, K.J., W.H. Desvousges, F.R. Johnson, R.W. Dunford and S.P. Hudson (1994), 'An investigation of part-whole biases in contingent valuation studies', *Journal of Environmental Economics and Management*, 27, 64–83.

Boxall, P.C., W.L. Adamowicz, J. Swait, M. Williams and J. Louviere (1996), 'A comparison of stated preference methods for environmental valuation', *Ecological Economics*, 18, 243–53.

Brookshire, D.S., B.C. Ives and W.D. Schulze (1976), 'The valuation of aesthetic preferences', *Journal of Environmental Economics and Management*, 3, 325–46.

Brookshire, D.S., R.C. d'Arge, W.D. Schulze and M.A. Thayer (1981), 'Experiments in valuing public goods', in V.K. Smith (ed.), *Advances in Applied Microeconomics: A Research Annual*, vol. 1, Greenwich, CT: JAI Press Inc.

Brouwer, R., N.A. Powe, R.K. Turner, I.H. Langford and I.J. Bateman (1999), 'Public attitudes to contingent valuation and public consultation', *Environmental Values*, 8, 325–47.

Brown, T.C., G.L. Peterson and B.E. Tonn (1995), 'The values jury to aid natural resource decisions', *Land Economics*, **71**(2), 250–60.

Bryman, A. (2004), *Social Research Methods*, 2nd edn, Oxford: Oxford University Press.

Bryman, A. and R.G. Burgess (eds) (1994), *Analysing Qualitative Data* London: Routledge.

Burgess, J. (1996), 'Focusing on fear: the use of focus groups in a project for the Community Forest Unit, Countryside Commission', *Area*, **28**(22), 130–35.

Burgess, J., J. Clark and C.M. Harrison (1998), 'Respondents' evaluations of a CV survey: a case study based on an economic valuation of the Wildlife Enhancement Scheme, Pevensey Levels in East Sussex', *Area*, **30**(1), 19–27.

Burgess, J., J. Clark and C.M. Harrison (2000), 'Culture, communication, and the information problem in contingent valuation surveys: a case study of a Wildlife Enhancement Scheme', *Environment and Planning C, Government and Policy*, **18**(5), 505–24.

Burgess, J., C. Harrison and M. Limb (1988a), 'Exploring environmental values through the medium of small groups. Part 1: theory and practice', *Environment and Planning A*, 20, 309–26.

Burgess, J., C. Harrison and M. Limb (1988b), 'Exploring environmental values through the medium of small groups. Part 2: illustrations of a group at work' *Environment and Planning A*, 20, 457–76.

Calia, P. and E. Strazzera (2000), 'Bias and efficiency of single versus double bound models for contingent valuation studies: a Monte Carlo analysis', *Applied Economics*, 32, 1329–36.

Cameron, J.I. (1997), 'Applying socio-ecological economics: a case study of contingent valuation and integrated catchment management', *Ecological Economics*, 23, 155–65.

Cameron, T.A. and J. Quiggin (1994), 'Estimation using contingent valuation data from a "dichotomous choice with follow-up" questionnaire', *Journal of Environmental Economics and Management*, 27, 218–34.

Carson, R.T. (1997), 'Contingent valuation surveys and tests of insensitivity to scope', in R.J. Kopp, W.W. Pommerehne and N. Schwarz (eds), *Determining the Value of Non-market Goods: Economic, Psychological and Policy Relevant Aspects of Contingent Valuation Methods,* London: Kluwer Academic Publishers.

Carson R.T. and R.M. Mitchell (1993), 'The value of clean water: the public's willingness to pay for boatable, fishable and swimable quality water', *Water Resources Research*, 29, 2445–54.

Carson, R.T. and R.M. Mitchell (1995), 'Sequencing and nesting in contingent valuation surveys', *Journal of Environmental Economics and Management*, **28**(2), 155–73.

Carson, R.T., N.E. Flores and W.M. Hanemann (1998), 'Sequencing and valuing public goods', *Journal of Environmental Economics and Management*, **36**(3), 314–23.

Carson, R.T., N.E. Flores and N.F. Meade (2001), 'Contingent valuation: controversies and evidence', *Environmental and Resource Economics*, 19, 173–210.

Carson, R.T., T. Groves and M.J. Machina (1999), 'Incentive and information properties of preference questions', paper presented at the European Association of Resource and Environmental Economics conference, Oslo.

Carson, R.T., M. Mitchell, M. Hanemann, R. Kopp, S. Presser and P. Ruud (1992), 'A contingent valuation study of lost passive use values resulting from the Exxon Valdez oil spill', report to the Attorney General of Alaska, commissioned by the State of Alaska, Juneau, AL.

Carson, R.T., L. Wilks and D. Imber (1994), 'Valuing the preservation of Australia's Kakadu conservation zone', *Oxford Economic Papers*, 46, 721–49.

Chess, C. and K. Purcell (1999), 'Public participation and the environment: do we know what works?', *Environmental Science & Technology*, **33**(16), 2885–692.

Chilton, S.M. and W.G. Hutchinson (1999), 'Exploring divergence between respondent and researcher definitions of the good in contingent valuation studies', *Journal of Agricultural Economics*, **50**(1), 1–16.

Chilton, S.M. and W.G. Hutchinson (2003), 'A qualitative examination of how respondents in a contingent valuation study rationalise their WTP responses to an increase in the quantity of the environmental good', *Journal of Economic Perspectives*, 24, 65–75.

Christie, M., N. Hanley, J. Warren, K. Murphy, R. Wright and T. Hyde (2006), 'Valuing the diversity of biodiversity', *Ecological Economics*, 58, 304–17.

Clark, J., J. Burgess and C.M. Harrison (2000), '"I struggled with this money business": respondents' perspectives on contingent valuation', *Ecological Economics*, 33, 45–62.

Clarke, P.M. (2000), 'Valuing the benefits of mobile mammographic screening units using the contingent valuation method', *Applied Economics*, 32, 1647–55.

Coffey, A. and P. Atkinson (1994), *Making Sense of Qualitative Data: Complementary Strategies*, Thousand Oaks, CA: Sage.

Coombs, C. (1964), *A Theory of Data*, New York: Wiley.

Cooper, J.C. and M.W. Hanemann (1995), 'Referendum contingent valuation: how many bounds are enough?', Food and Agricultural Organisation Economic and Social Department, working paper, Italy.

Coote, A. and J. Lenaghan (1997), *Citizens Juries: Theory into Practice*, London: Institute for Public Policy Research.

Crosby, N., J.M. Kelly and P. Schaefer (1986), 'Citizens panels: a new approach to citizen participation', *Public Administration Review*, **46**(2), 170–78.

Cummings, R.G. and L.O. Taylor (1999), 'Unbiased value estimates for environmental goods: cheap talk design for the contingent valuation method', *American Economic Review*, **89**(3), 649–65

Cummings, R.G., D.S. Brookshire and W.D. Schulze (1986), *Valuing Environmental Goods: An Assessment of the Contingent Valuation Method*, Totowa, NJ: Rowman & Allanheld.

Curtis, I.A. (2004), 'Valuing ecosystem goods and services: a new approach using a surrogate market and the combination of a multiple criteria analysis and a Dephi panel to assign weights to the attributes', *Ecological Economics*, 50, 163–94.

Davies, A.R. (1999), 'Where do we go from here? Environmental focus groups and planning policy formation', *Local Environment*, **4**(3), 297–316.

De Jong, M. and P.J. Schellens (1998), 'Focus groups or individual interview? A comparison of text evaluation approaches', *Technical Communication*, 45, 77–88.

Department of Health (1999), *Economic Appraisal of the Health Effects of Air Pollution*, Ad-Hoc Group on the Economic Appraisal of the Health Effects of Air Pollution, London: The Stationery Office.

DeShazo, J.R. (2002), 'Designing transactions without framing effects in iterative question formats', *Journal of Environmental Economics and Management*, 43, 360–85.

Desvousges, W.H., V.K. Smith, D.J. Brown and D.K. Pate (1984), *The Role of Focus Groups in Designing a Contingent Valuation Survey to Measure the Benefits of Hazardous Waste Management Regulation*, RTI project no. 2505–13 Chapel Hill, NC: Research Triangle Institute.

Diamond, P.A. and J.A. Hausman (1993), 'On contingent valuation measurement of nonuse value', in J.A. Hausman (eds), *Contingent Valuation: A Critical Assessment*, Amsterdam: North Holland.

Diamond, P.A. and J.A. Hausman (1994), 'Contingent valuation: is some number better than no number?', *Journal of Economic Perspectives*, **8**(4), 45–64.

Dienel, P. (1997), *Die Plannungszelle: der Burger plant seine unwelt. Eine alternative zur establishment-demokratie*, Opladen: Westdeutscher Verlage.

Dryzek, J.S. (1990), *Discursive Democracy: Politics, Policy and Political Science*, New York: Cambridge University Press.

Dryzek, J.S. (2000), *Deliberative Democracy and Beyond: Liberals, Critics, Contestations*, Oxford: Oxford University Press.

Elster, J. (1983), *Sour Grapes, Studies in the Subversion of Rationality*, Cambridge: Cambridge University Press.

Elster, J. (1998), 'Introduction', in J. Elster (ed.), *Deliberative Democracy*, Cambridge: Cambridge University Press.

Farber, S.C., R. Costanza and M.A. Wilson (2002), 'Economic and ecological concepts for valuing ecosystem services', *Ecological Economics*, 41, 375–92.

Fischhoff, B. (1991), 'Value elicitation: is there anything in there?' *American Psychologist*, 46(8), 835–47.

Fischhoff, B. (1997), 'What do psychologists want? Contingent valuation as a special case of asking questions', in R.J. Kopp, W.W. Pommerehne and N. Schwarz (eds), *Determining the Value of Non-Market Goods: Economic, Psychological, and Policy Relevant Aspects of Contingent Valuation*, London: Kluwer Academic Publishers.

Fischhoff, B. and L. Furby (1988), 'Measuring values: a conceptual framework for interpreting transactions', *Journal of Risk and Uncertainty*, 1, 147–84.

Fischhoff, B., M.J. Quadrel, M. Kamlet, G. Loewenstein, R. Dawes, P. Fischbeck, S. Klepper, J. Leland and P. Stroh (1993), 'Embedding effects: stimulus representation and response mode', *Journal of Risk and Uncertainty*, 6, 211–34.

Fischhoff, B., P. Slovic and S. Lichtenstein (1978), 'Fault trees: sensitivity of estimated failure probabilities to problem representation', *Journal of Experimental Psychology: Human Perception and Performance*, 4, 330–40.

Fischhoff, B., N. Welch and S. Frederick (1999), 'Construal processes in preferences assessment', *Journal of Risk and Uncertainty*, **19**(1/3), 139–65.

Fishbein, M. and I. Ajzen (1975), *Belief, Attitude, Intention and Behaviour: An Introduction to Theory and Research*, Reading, MA: Addison-Wesley.

Fisher, A.C. (1996), 'The conceptual underpinnings of the contingent valuation method', in D.J. Bjornstad and J.R. Kahn (eds), *The Contingent Valuation of Environmental Resources: Methodological Issues and Research Needs*, Cheltenham, UK and Brookfield, VT, US:Edward Elgar.

Foster, V. and S. Mourato (2000), 'Valuing the multiple impacts of pesticide use in the UK: a contingent ranking approach', *Journal of Agricultural Economics*, **51**(1), 1–21.

Foster, V. and S. Mourato (2002), 'Testing for consistency in contingent ranking experiments', *Journal of Environmental Economics and Management*, 44, 309–28.

Frey, B. (1997), 'A constitution for knaves crowds out civic virtues', *The Economic Journal*, 107, 1043–53.

Frykblom, P. (1997), 'Hypothetical question modes and real willingness to pay', *Journal of Environmental Economics and Management*, 34, 275–87.

Frykblom, P. and Shogren, J. (2000), 'An experimental testing of anchoring effects in discrete choice questions', *Environmental and Resource Economics*, **16**(3), 329–341.

Garrod, G.D. and K.G. Willis (1997), 'The non-market benefits of enhancing forest biodiversity: a contingent ranking study', *Ecological Economics*, 21, 45–61.

Garrod, G.D. and K.G. Willis (1999), *Economic Valuation of the Environment: Methods and Case Studies*, Cheltenham, UK and Northampton, MA, US: Edward Elgar.

Goldman, A.E. and S.S. McDonald (1987), *The Group Depth Interview: Principles and Practice*, Englewood Cliffs, NJ: Prentice Hall.

Goodin, R. (1986), 'Laundering preferences', in J. Elster and A. Hylland (eds), *Foundations of Social Choice Theory*, Cambridge: Cambridge University Press.

Goodin, R.E. (1992), *Motivating Political Morality*, Oxford: Basil Blackwell.

Green, C.H. and S.M. Tunstall (1999), 'A psychological perspective', in I.J. Bateman and K.G. Willis (eds), *Valuing Environmental Preferences: Theory and Practice of the Contingent Valuation Method in the US, EU and Developing Countries*, Oxford: Oxford University Press.

Gregory, R. and P. Slovic (1997), 'A constructive approach to environmental valuation', *Ecological Economics*, **21**(3), 175–81.

Gregory, R., S. Lichtenstein and P. Slovic (1993), 'Valuing environmental resources: a constructive approach', *Journal of Risk and Uncertainty*, 7, 177–97.

Gregory, R. and K. Wellman (2001), 'Bringing stakeholder values into environmental policy choices: a community-based estuary case study', *Ecological Economics*, 39, 37–52.

Gundry, K.G. and T.A. Heberlein (1984), 'Do public meetings represent the public?', *Journal of the American Planning Association*, Spring, 175–82.

Habermas, J. (1984), *The Theory of Communicative Action*, Boston, MA: Beacon Press.

Hammack, J. and G.M. Brown, Jr. (1974), *Waterfowl and Wetlands: Towards Bioeconomic Analysis*, Washington, DC: Resources for the Future.

Hammitt, J.K. and J.D. Graham (1999), 'Willingness to pay for health protection: inadequate sensitivity to probability', *Journal of Risk and Uncertainty*, **8**(1), 33–62.

Hanemann, W.M. (1994), 'Valuing the environment through contingent valuation', *Journal of Economic Perspectives*, **8**(4), 19–43.

Hanemann W.M. and Kanninen B. (1999), 'The statistical analysis of discrete-response CV data', in I.J. Bateman and K.G. Willis (eds), *Valuing Environmental Preferences: Theory and Practice of the Contingent Valuation Method in the US, EU and Developing Countries*, Oxford: Oxford University Press.

Hanemann, W.M., J. Loomis and B. Kanninen (1991), 'Statistical efficiency of double-bounded dichotomous choice contingent valuation', *American Journal of Agricultural Economics* **73**(4), 1255–63.

Hanley, N. (2001), 'Cost-benefit analysis and environmental policymaking', *Environment and Planning C: Government and Policy*, 19, 103–18.

Hanley, N. and C.L. Spash (1993), *Cost-benefit Analysis and the Environment*, Cheltenham, UK and Brookfield, VT, US: Edward Elgar.

Hanley, N., D. Macmillan, R.E. Wright, C. Bulluck, I. Simpson, D. Parsisson and B. Crabtree (1998), 'Contingent valuation versus choice experiments: estimating the benefits of Environmentally Sensitive Areas in Scotland', *Journal of Agricultural Economics*, **49**(1), 1–15.

Hanley, N., J. Shogren and B. White (2001), *Introduction to Environmental Economics*, Oxford: Oxford University Press.

Hanley, N., K.G. Willis, N.A. Powe and M. Anderson (2002), 'Valuing the benefits of biodiversity in forests', report to the Forestry Commission, accessed at www.forestry.gov.uk

Hausman J.A. (ed.) (1993), *Contingent Valuation: A Critical Assessment*, Amsterdam: North Holland.

Hausman, J.A. and P.A. Ruud (1987), 'Specifying and testing econometric models for rank-ordered data', *Journal of Econometrics*, 34, 83–104.

Hoehn, J.P. (1991), 'Valuing the multidimensional impacts of environmental policy: theory and methods', *American Journal of Agricultural Economics*, **73**(2), 289–99.

Hoehn, J.P. and A. Randall (1987), 'A satisfactory benefit cost indicator from contingent valuation', *Journal of Environmental Economics and Management*, 14, 226–47.

Hoehn, J.P. and A. Randall (1989), 'Too many proposals pass the benefit cost test', *American Economic Review*, **79**(3), 544–51.

Hoevenagel, R. (1994), 'The contingent valuation method: scope and validity', PhD dissertation for the Vrije Unversiteit, Amsterdam.

Hoevenagel, R. (1996), 'The Validity of the contingent valuation method: perfect and regular embedding', *Environmental and Resource Economics*, 7, 57–78.

Horowitz, J. and K. McConnell (2002), 'A review of WTA-WTP studies', *Journal of Environmental Economics and Management*, **44**(3), 426–47.

Howarth, R.B. and M.A. Wilson (2006), 'A theoretical approach to deliberative valuation: aggregation by mutual consent', *Land Economics*, **82**(1), 1–16.

Howe, C.W. and M.G. Smith (1994), 'The value of water supply reliability in urban water systems', *Journal of Environmental Economics and Management*, 26, 19–30.

Irwin, J.R., P. Slovic, S. Lichtenstein and G.H. McClelland (1993), 'Preference reversals and the measurement of environmental values', *Journal of Risk and Uncertainty*, 6, 5–18.

Jacobs, M. (1997), 'Environmental valuation, deliberative democracy and public decision making institutions', in J. Foster (ed.), *Valuing Nature? Economics, Ethics and Environment*, London: Routledge, Chapter 13.

James, R.F. and R.K. Blamey (2005), 'Deliberation and economic valuation', in M. Getzner, C.L. Spash and S. Stagl (eds), *Alternatives for Environmental Valuation, Routledge Explorations in Environmental Economics*, London: Routledge.

Janis, I.L. (1982), *Groupthink*, Boston, MA: Houghton Mifflin.

Johannesson, M., B. Liljas and R. O'Connor (1997), 'Hypothetical versus real willingness to pay: some experimental results', *Appied Economics Letters*, **4**(3), 149–51.

Johansson, P.O., B. Kriström and K.G. Mäler (1989), 'Welfare evaluations in contingent valuation experiments with discrete response data: comment', *American Journal of Agricultural Economics*, 71, 1054–56.

Johnson, K.N., R.L. Johnson, D.K. Edwards and C.A. Wheaton (1993), 'Public participation in wildlife management: opinions from public meetings and random surveys', *Wildlife Society Bulletin*, 21, 218–25.

Johnston, R.J. (2006), 'Is hypothetical bias universal? Validating contingent valuation responses using a binding public referendum', *Journal of Environmental Economics and Management*, 52, 469–481.

Johnston, R.J., T.F. Weaver, L.A. Smith and S.K. Swallow (1995), 'Contingent valuation focus groups: insights from ethnographic interview techniques', *Agricultural and Resource Economics Review*, **24**(1), 56–69.

Jourard, S.M. (1964), *The Transparent Self*, Princeton, NJ: Van Nostrand.

Kahneman, D. (1986), Comments within Chapter 12 of R.G. Cummings, D.S. Brookshire and W.D. Schulze (eds), *Valuing Environmental Goods: An Assessment of the Contingent Valuation Method*, Totowa, NJ: Rowman and Allanheld.

Kahneman, D. and Knetsch, J.L. (1992), 'Valuing public goods: the purchase of moral satisfaction', *Journal of Environmental Economics and Management*, 22, 55–70.

Kahneman, D., I. Ritov and D. Schkade (1999), 'Economic preferences or attitude expressions?: An analysis of dollar response to public issues', *Journal of Risk and Uncertainty*, **19**(1–3), 203–35.

Kaplowitz, M.D. (2000), 'Identifying ecosystem services using multiple methods: lessons from the mangrove wetlands of Yucatan, Mexico', *Agriculture and Human Values*, 17, 169–79.

Kaplowitz, M.D. and Hoehn, J.P. (2001), 'Do focus groups and individual interviews reveal the same information for natural resource valuation?', *Ecological Economics*, 36, 237–47.

Kealy, M.J. and R.W. Turner (1993), 'Testing of equality of contingent valuations', *American Journal of Agricultural Economics*, **75**(2), 321–31.

Kealy, M.J., M. Montgomery and J.F. Dovidio (1990), 'Reliability and predictive validity of contingent valuation values: does the nature of the good matter?', *Journal of Environmental Valuation and Management*, 19, 244–263.

Keeney, R.L. (1980), *Siting Energy Facilities*, New York: Academic.

Kenyon, W. and N. Hanley (2005), 'Three approaches to valuing nature, forest floodplain restoration', in M. Getzner, C.L. Spash and S. Stagl (eds), *Alternatives for Environmental Valuation*, Routledge Explorations in Environmental Economics, London: Routledge.

Kenyon, W., N. Hanley and C. Nevin (2001), 'Citizens' juries: an aid to environmental valuation?' *Environment and Planning C: Government and Policy*, 19, 557–66.

Kenyon, W. and C. Nevin (2001), 'The use of economic and participatory approaches to assess forest development: a case study in the Ettick Valley', *Forest Policy and Economics*, 3, 69–80.

Kreps, D.M. (1990), *A Course in Microeconomic Theory*, London: Harvester Wheatsheaf.

Krueger, R.A. (1993), 'Quality control in focus group research', in D.L. Morgan (eds), *Successful Focus Groups: Advancing the State of the Art,* Newbury Park, CA: Sage Publications.

Krueger, R.A. (1994), *Focus Groups: A Practical Guide for Applied Research*, Thousand Oaks, CA: Sage Publications.

Krueger, R.A. and M.A. Casey (2000), *Focus Groups: A Practical Guide for Applied Research*, London: Sage Publications.

Krutilla, J.V. (1967), 'Conservation reconsidered', *The American Economic Review*, 57, 777–86.

Kwak, S.-J., S.-H. Yoo and T.-Y. Kim (2001), 'A constructive approach to air-quality valuation in Korea', *Ecological Economics*, 38, 327–44.

Larson, D.M. (1993), 'On measuring existence value', *Land Economics*, **69**(4), 377–88.

Lazo, J.K., W.D. Schulze, G.H. McClelland and J.K. Doyle (1992), 'Can contingent valuation measure nonuse values?', *American Journal of Agricultural Economics*, 74, 1126–32.

Levine, J.M. and R.L. Moreland (1995), 'Group processes', in J.M. Tesser (ed.), *Advanced Social Psychology*, New York: McGraw Hill.

Lichtenstein, S. and P. Slovic (1973), 'Reversals of preference between bids and choices in gambling decisions', *Journal of Experimental Psychology*, 89, 46–55.

Lienhoop, N. and D. Macmillan (2006), 'Valuing wilderness in Iceland: estimation of WTA and WTP using the market stall approach to contingent valuation', *Land Use Policy*, **24**(1), 289–95.

Lienhoop, N. and D. Macmillan (2007), 'Valuing a complex environmental change: assessing participant performance in deliberative group-based approaches and in-person interviews for contingent valuation', *Environmental Values*, forthcoming.

Lockwood, M. (1998), 'Integrated value assessment using paired comparisons', *Ecological Economics*, **25**(1), 83–93.

Loomis, J., M. Lockwood and T. DeLacy (1993), 'Some empirical evidence on embedding effects in contingent valuation of forest protection', *Journal of Environmental Economics and Management*, 24, 45–55.

MacLean, D. (1991), 'A critical look at informed consent', unpublished manuscript, University of Maryland at Baltimore County, Catonsville, MD.

Macmillan, D.C., L. Philip, N. Hanley and B. Alvarez-Farizo (2002), 'Valuing the non-market benefits of wild goose conservation: a comparison of interview and group-based approaches', *Ecological Economics*, 43, 49–59.

Macmillan, D., N. Hanley and N. Lienhoop (2006), 'Contingent valuation: Environmental polling or preference engine?' *Ecological Economics*, **60**(1), 299–307.

Margolis, H. (1982), *Selfishness, Altruism, and Rationality*, Cambridge: Cambridge University Press.

McCelland, E. (2001), *Measurement Issues and Validity Tests for Using Attitude Indicators in Contingent Valuation Research*, Washington, DC: US Environmental Protection Agency.

McComas, K.A. (2001), 'Public meetings about local waste management problems, comparing participants to nonparticipants', *Environmental Management*, **27**(1), 135–47.

McComas, K.A. and C.W. Scherer (1998), 'Reassessing public meetings as participation in risk management decisions', *Risk: Health, Safety and Environment*, **9**(4), 347–60.

McDaniels, T.L. and C. Roessler (1998), 'Multiattribute elicitation of wilderness preservation benefits: a constructive approach', *Ecological Economics*, 27, 299–312.

McDaniels, R.L., R. Gregory, J. Arvai and R. Chuenpagdee (2003), 'Decision structuring to alleviate embedding in environmental valuation', *Ecological Economics*, 46, 33–46.

McFadden, D. (1973), 'On conditional logit analysis of qualitative choice behavior', in P. Zarembka (ed.), *Frontiers of Econometrics*, New York: Academic Press.

McFadden, D. and G. Leonard (1993), 'Issues in the contingent valuation of environmental goods: methodologies for data collection and analysis', in J. A. Hausman (ed.), *Contingent Valuation: A Critical Assessment*, Amsterdam: North-Holland.

Merton, R.K. and P.L. Kendal (1946), 'The focused interview', *American Journal of Sociology*, 51, 541–57.

Merton, R.K. (1987), 'The focussed interview and focus groups: continuities and discontinuities', *Public Opinion Quarterly*, 51, 550–66.

Merton, R.K., M. Fiske and P.L. Kendall (1990), *The Focused Interview*, 2nd edn, New York: Free Press.

Miles, M.B. and A.M. Huberman (1994), *Qualitative Data Analysis*, 2nd edn, Thousand Oaks, CA: Sage Publications.

Mitchell, R.C and R.T. Carson (1989), *Using Surveys to Value Public Goods: The Contingent Valuation Method*, Washington DC: Resources for the Future.

Morgan, D.L. (1997), *Focus Groups as Qualitative Research*, Qualitative Research Methods Series, 16, 2nd edn, London: Sage Publications Ltd.

Morgan, D.L. and R.A. Krueger (1993), 'When to use focus groups and why', in D.L. Morgan (ed.), *Successful Focus Groups-Advancing the State of the Art*, Newbury Park, CA: Sage Publications.

Morgan, D.L. and M.T. Spanish (1985), 'Social interaction and the cognitive organisation of health-relevant behaviour', *Sociology and Health of Illness*, 7, 401–22.

Navrud, S. and R. Ready (eds) (2006), *Environmental Value Transfer: Issues and Methods, The Economics of Non-Market Goods and Resources*, vol. 9, London: Springer.

Neill, H.R. (1995), 'The context for substitutes in CVM studies: some empirical observations', *Journal of Environmental Economics and Management*, 29, 393–7.

Neill, H.R., R.G. Cummings, P.T. Ganderton, G.W. Harrison and T. McGuckin (1994), 'Hypothetical surveys and real economic commitments', *Land Economics*, **70**(2), 145–54.

Nicholson, W. (1989), *Microeconomic Theory: Basic Principles and Extensions*. 4th edn, Chicago, IL: Dryden Press.

Niemeyer, S. and C.L. Spash (2001), 'Environmental valuation analysis, public deliberation, and their pragmatic syntheses: a critical appraisal', *Environment and Planning C: Government & Policy*, **19**(4), 567–86.

Nunes, P.A.L.D. (2002), 'Using factor analysis to identify consumer preferences for the protection of a natural area in Portugal', *European Journal of Operational Research*, 140, 499–516.

Nunes, P.A.L.D. and E. Schokkaert (2003), 'Identifying the warm glow effect in contingent valuation', *Journal of Environmental Economics and Management*, 45, 231–45.

O'Neil, J. (1997), 'Value pluralism, incommensurability and institutions', in J. Foster (ed.), *Valuing Nature? Economics, Ethics and Environment*, London: Routledge, Chapter 5.

O'Riordan, T. (1985), 'Future directions in environmental policy', *Environment and Planning A*, 17, 1431–46.

Opaluch, J.J., S. Swallow, T. Weaver, C. Wessels and D. Wichlens (1993), 'Evaluating impacts from noxious waste facilities, including public preferences in current siting mechanisms', *Journal of Environmental Economics and Management,* 24, 41–59.

Oppenheim, A.N. (1992), *Questionnaire Design, Interviewing and Attitude Measurement*, New Edition: London, Continuum.

Ostrom, E. (2000), 'Collective action and the evolution of social norms', *Journal of Economic Perspectives*, **14**(3), 137–58.

Payne, J.W., J.R. Bettman and E.J. Johnson (1992), 'Behavioral decision research: a constructive processing perspective', *Annual Review of Psychology*, 43, 87–131.

Pearce, D.W. (1983), *Cost-Benefit Analysis*, 2nd edn, London: Macmillan.

Petty, J., I. Guijit, J. Thompson and I. Scoones (1995), *Participatory Learning and Action: A Trainers Guide*, London: International Institute for Environment and Development.

Philip, L.J. and D.C. Macmillan (2005), 'Exploring values, context and perceptions in contingent valuation studies: the CV Market Stall technique and willingness to pay for wildlife conservation', *Journal of Environmental Planning and Management*, **48**(2), 257–74.

Powe, N.A. (2000), 'Using contingent valuation to value nested goods: a case of the Broadland Flood Alleviation Scheme', PhD thesis for the School of Environmental Sciences, University of East Anglia, Norwich.

Powe, N.A. and I.J. Bateman (2003), 'Ordering effects in nested "top-down" and "bottom-up" contingent valuation designs', *Ecological Economics*, **45**(2), 255–70.

Powe, N.A. and Bateman, I.J. (2004), 'Investigating insensitivity to scope: a split-sample test of perceived scheme realism', *Land Economics*, **80**(2), 258–71.

Powe, N.A., W.A. Wadsworth, G.D. Garrod and P.L. McMahon (2004a), 'Putting action into biodiversity planning: assessing preferences towards funding', *Journal of Environmental Planning and Management*, **47**(2), 287–301.

Powe, N.A., G.D. Garrod, P.L. McMahon and K.G. Willis (2004b), 'Assessing customer preferences for water supply options using mixed methodology choice experiments', *Water Policy*, 6, 427–41.

Powe, N.A., G.D. Garrod and P.L. McMahon (2005), 'Mixing methods within stated preference environmental valuation: choice experiments and post-questionnaire qualitative analysis', *Ecological Economics*, **52**(4), 513–26.

Powe, N.A., K.G. Willis and G.D. Garrod (2006), 'Difficulties in valuing street light improvement: trust, surprise and bound effects', *Applied Economics*, **38**(4), 371–83.

Randall, A (1991), 'Total and nonuse values', in J.B. Braden and C.D. Kolstad

(eds), *Measuring the Demand for Environmental Quality*, London: North-Holland.
Rice, S.A. (ed.) (1931), *Methods in Social Science*, Chicago, IL: University of Chicago Press.
Rokeach, M. (1973), *The Nature of Human Values*, New York: Free Press.
Rowe, R., R. d'Arge and D.S. Brookshire (1980), 'An experiment on the economic value of visibility', *Journal of Environmental Economics and Management* 7, 1–19.
Rawls, J. (1974), 'Some reasons for the maximum criterion', *American Economic Review*, **64**(2), 141–6.
Sagoff, M. (1988), *The Economy of the Earth*, Cambridge: Cambridge University Press.
Sagoff, M. (1998), 'Aggregation and deliberation in valuing environmental public goods: a look beyond contingent pricing', *Ecological Economics*, **24**(2–3), 213–31.
Samuelson, P. (1954), 'The pure theory of public expenditure', *Review of Economics and Statistics*, 36, 387–9.
Scarpa, R. and I.J. Bateman (2000), 'Efficiency gains afforded by improved bid design versus follow-up valuation questions in discrete-choice CV studies', *Land Economics*, **76**(2), 299–311.
Schkade, D.A. and J.W. Payne (1993), 'Where do the numbers come from? How people respond to contingent valuation questions', in J.A. Hausman (ed.), *Contingent Valuation: A Critical Assessment*, Amsterdam: North Holland.
Schkade, D.A. and J.W. Payne (1994), 'How people respond to contingent valuation questions: a verbal protocol analysis of willingness to pay for an environmental regulation', *Journal of Environmental Economics and Management*, 26, 88–109.
Schulze, W.D., R.C. d'Arge and D.S. Brookshire (1981), 'Valuing environmental commodities: some recent experiments', *Land Economics*, **57**(2), 151–72.
Schulze, W.D., G.H. McClelland, J.K. Lazo and R.D. Rowe (1998), 'Embedding and calibration in measuring non-use values', *Resource and Energy Economics*, **20**(2), 163–78.
Schwartz, S.H. (1977), 'Normative influences on altruism', in L. Berkowitz (ed.), *Advances in Experimental Social Psychology*, vol. 10, New York: Academic Press.
Simon, H. (1956), 'Rational choice and the structure of the environment', *Psychological Review*, 63, 129–38.
Sinclair, M. (1977), 'The public hearing as a participatory device: evaluation of the IJC experience', in W.R.D. Sewell and J.T. Coppock (eds), *Regulating Risk: The Science and Politics of Risk,* Washington, DC: International Life Sciences Institute.

Slovic, P. (1995), 'The construction of preference', *American Psychologist*, **50**(5), 364–71.

Smith, G. (2003), 'Deliberative democracy and the environment', in *Routledge Research in Environmental Politics*, London: Routledge.

Smith, V.K. (1992), 'Arbitrary values, good causes, and premature verdicts: comment', *Journal of Environmental Economics and Management*, 22, 71–89.

Smith, V.K. and L.L. Osborne (1996), 'Do contingent valuation estimates pass a "scope" test? A meta analysis', *Journal of Environmental Economics and Management*, **31**(3), 287–301.

Spash, C.L. (2000), 'Ecosystem, contingent valuation and ethics: the case of wetland re-creation', *Ecological Economics*, 34, 195–215.

Spash, C.L. and N. Hanley (1995), 'Preferences, information and biodiversity preservation', *Ecological Economics*, 12, 191–208.

Stevens, T.H., J. Echeverria, R.J. Glass, T. Hager and T.A. More (1991), 'Measuring the existence value of wildlife: what do CVM estimates really show', *Land Economics*, **67**(4), 390–400.

Stevens, T.H., R.A. More and R.J. Glass (1994), 'Interpretation and temporal stability of CV bids for wildlife existence: a panel study', *Land Economics*, **70**(3), 355–63.

Strack, F. and N. Schwarz (1992), 'Communicative influences in standardized question situations', in G.R. Semin and K. Fielder (eds), *Language Interaction and Social Cognition*, London: Sage.

Strauss, A.L. and J. Corbin (1998), *Basics of Qualitative Research : Techniques and Procedures for Developing Grounded Theory*, Thousand Oaks, CA: Sage Publications.

Sussman, S., D. Burton, C.W. Dent, A.W. Stacy and B.R. Flay (1991), 'Use of focus groups in developing an adolescent tobacco use cessation program: collective norm effects', *Journal of Applied Social Psychology*, 21, 1772–82.

Svedsäter, H. (2003), 'Economic valuation of the environment: how citizens make sense of contingent valuation questions', *Land Economics*, **79**(1), 122–35.

Turner, R.K. and T. Jones (eds) (1991), *Wetlands, Market and Intervention Failures*, London: Earthscan.

Turner, R.K., D. Dent and R.D. Hey (1983), 'Valuation of the environmental impact of wetland flood protection and drainage schemes', *Environment and Planning A*, 15, 871–88.

Turner, R.K., J.C.J.M. van den Bergh, T. Söderqvist, A. Barendregt, J. van der Straaten, E. Maltby and van E.C. Ierland (2000), 'Ecological-economic analysis of wetlands: scientific integration for management and policy', *Ecological Economics*, 35, 7–25.

Turner, R.K., J.M. van den Bergh and R. Brouwer (2003), 'Introduction', in R.K. Turner, J.M. van den Bergh and R. Brouwer (eds), *Managing Wetlands: An Ecological Economics Approach*, Cheltenham, UK and Northampton, MA, US: Edward Elgar.

Tversky, A. and D. Kahneman (1974), 'Judgement under uncertainty: heuristics and biases', *Science*, 185, 1124–31.

Tversky, A., P. Slovic and S. Sattath (1988), 'Contingent weighting in judgement and choice', *Psychological Review*, **95**(3), 371–84.

United Nations (1993), *The Global Partnership for the Environment and Development. A Guide to Agenda 21*, post-Rio edn, New York: United Nations.

Vadnjal, D. and M. O'Connor (1994), 'What is the value of Rangitoto Island?', *Environmental Values*, 3, 369–80.

Varian H.R. (1992), *Microeconomic Analysis*, 3rd edn, London: W.W. Norton & Company.

Varian, H.R. (1999), 'Commentary on "economic preferences or attitude expressions: an analysis of dollar responses to public issues" by Kahneman *et al.*', *Journal of Risk and Uncertainty*, **19**(1–3), 241–2.

Vatn, A. (2004), 'Environmental valuation and rationality', *Land Economics*, **80**(1), 1–18.

Vatn, A. (2005), *Institutions and the Environment*, Cheltenham, and Northampton, MA, US: Edward Elgar.

Vatn, A. and D.W. Bromley (1994), 'Choices without prices without apologies', *Journal of Environmental Economics and Management*, 26, 129–48.

Viscusi, W.K., W.A. Megat and J. Huber (1991), 'Pricing environmental health risks: survey assessments for risk-risk and risk-dollar trade-offs for chronic bronchitis', *Journal of Environmental Economics and Management,* 21, 32–51.

Ward, H. (1999), 'Citizens' juries and valuing the environment: a proposal', *Environmental Politics*, **8**(2), 75–96.

Welch, N. and B. Fischhoff (2001), 'The social context of contingent valuation transactions', *Society and Natural Resources*, **14**(3), 209–21.

Welsh, M. and G.L. Poe (1998), 'Elicitation effects in contingent valuation: comparisons to a multiple-bounded discrete choice approach', *Journal of Environmental Economics and Management*, 34, 219–32.

Whitehead, J.C. and G.C. Blomquist (1991), 'Measuring contingent values for wetlands: effects of information about related environmental goods', *Water Resources Research*, **27**(10), 2523–31.

Whittington, D., V.K. Smith, A. Okorafor, A. Okore, J.L. Liu and A. McPhail (1992), 'Giving respondents more time to think in contingent valuation studies: a developing country application', *Journal of Environmental Economics and Management*, 22, 205–25.

Willis, K.G., G.D. Garrod, J.F. Benson and M. Carter (1996), 'Benefits and costs of the wildlife enhancement scheme: a case study of the Pevensey Levels', *Journal of Environmental Planning and Management*, 39, 387–401.

Willis, K.G. and N.A. Powe (1998), 'Contingent valuation and real economic commitments: a private good experiment', *Journal of Environmental Planning and Management*, **41**(5), 611–19.

Willis, K.G., N.A. Powe and G.D. Garrod (2005), 'Estimating the value of improved street lighting: a factor analytical discrete choice approach', *Urban Studies*, **42**(12), 1–15.

Wilson, M.A. and R.B. Howarth (2002), 'Discourse-based valuation of ecosystem services: establishing fair outcomes through group deliberation', *Ecological Economics*, 41, 431–43.

von Winterfeldt, D. and W. Edwards (1986), *Decisions Analysis and Behavioral Research*, New York: Cambridge University Press.

Index